Compact Multifunctional Antennas for Wireless Systems

WILEY SERIES IN MICROWAVE AND OPTICAL ENGINEERING

KAI CHANG, Editor
Texas A&M University

A complete list of the titles in this series appears at the end of this volume.

Compact Multifunctional Antennas for Wireless Systems

ENG HOCK LIM

KWOK WA LEUNG

A JOHN WILEY & SONS, INC., PUBLICATION

Published by John Wiley & Sons, Inc., Hoboken, New Jersey.
Published simultaneously in Canada.

For general information on our other products and services or for technical support, please contact our
Customer Care Department within the United States at (800) 762-2974, outside the United States at
(317) 572-3993 or fax (317) 572-4002.

Wiley also publishes its books in a variety of electronic formats. Some content that appears in print
may not be available in electronic formats. For more information about Wiley products, visit our web
site at www.wiley.com.

Library of Congress Cataloging-in-Publication Data:

Lim, Eng Hock, 1974–
 Compact multifunctional antennas for wireless systems / Eng Hock Lim and Kwok Wa Leung.
  p. cm.
 Includes bibliographical references.
 ISBN 978-0-470-40732-5
 1. Antennas (Electronics) 2. Wireless communication systems–Equipment and supplies.
 I. Leung, K. W. (Kwok Wa), 1967– II. Title.
  TK7871.6.L56 2012
  621.384′135–dc23

                                                        2011040051

10 9 8 7 6 5 4 3 2 1

Contents

Preface

The objective of this book is to provide up-to-date information on modern multifunctional antennas and microwave circuits. Today, it is a trend to bundle multiple components into a single module to achieve high compactness and good signal quality. In the last two decades, the multifunctional concept has already been applied extensively to miniaturize various active and passive radio-frequency devices. Active antennas can be considered one of the earliest multifunctional antennas that have received a high level of attention from both academia and industry. Due to the rapid advancement of packaging technologies, various multifunctional devices can be made easily using such new techniques as antenna-on-package, antenna-in-package, and low-temperature-co-fired. Although there are many books describing the design of active and passive microwave systems, the multifunctional concept has yet to be fully explored for antennas and microwave circuits.

In this book, antennas are incorporated with active and passive microwave devices to design various multifunctional modules. The first part of the book introduces several novel passive components, such as an antenna filter and an antenna packaging cover. To make the coverage more complete, the development of the balun filter, a relatively new component, is also covered. Then, switches are integrated into antenna structures to achieve reconfiguration. Some recent work from the Institute of Applied Physics at the University of Electronic Science and Technology of China in Chengdu on frequency-, pattern-, and multireconfigurable antennas is discussed. Oscillating and amplifying antennas, which are among the conventional active antennas that have received much interest in recent decades, are featured in the book. Since the 1970s, oscillating antennas have been explored extensively as to power combining, phase locking, and beam switching. The reflection amplifier and coupled-load antenna oscillators are both visited and attention has been directed to their special applications. For example, it is shown that such active antennas can be made wearable as well as being used as a packaging cover. We focus on studying the receiving amplifying antennas, as the transmitting counterparts have been well explored in many other books. The co-design process of the amplifying antenna is discussed in detail. In the final part of the book, antennas are combined with solar cells to provide new applications. The design methods for various multifunctional antennas and microwave circuits are discussed, along with the elucidation of some important contemporary issues. We also explore the use of multiple software design tools in co-designing multifunctional antennas.

Acknowledgments

First, we would like to express our sincere gratitude to Professor Kai Chang (University of Texas, A&M) for his support of publishing this book. Special thanks go to Professor Kwai Man Luk for his kind encouragement of writing up this book. Another important person to whom we are thankful is Professor Quan Xue (City University of Hong Kong) for sharing his knowledge and experience in many discussions. We are particularly appreciative of the assistance provided by many colleagues at the State Key Laboratory of Millimeter Wave, City University of Hong Kong.

Our appreciation goes to Dr. Xue-Song Yang, Professor Shao-Qiu Xiao, and Professor Bing-Zhong Wang, all from the University of Electronic Science and Technology of China, for sharing their recent research work on reconfigurable antennas (Chapter 3). We would like to express many thanks to Professor Jian-Xin Chen (Nantung University, China), Dr. Jin Shi (I^2R, Singapore), Dr. Yong-Mei Pan (City University of Hong Kong), Dr. Shao-yong Zheng (City University of Hong Kong), and Dr. Kok Keong Chong (Universiti Tunku Abdul Rahman, Malaysia) for their help on countless occasions and their willingness to share much useful information.

Heartfelt gratitude to the following friends and students for their hard work in broadening the horizon of multifunctional antennas and microwave circuits: Xiao-Sheng Fang (City University of Hong Kong), Hong-Yik Wong (Universiti Tunku Abdul Rahman, Malaysia), Choon-Chung Su (Universiti Tunku Abdul Rahman, Malaysia), Chi-Hwa Ng (Agilent Technologies Sdn. Bhd., Malaysia), Gim-Hui Khor, and Kwan-Keen Chan.

Finally, we would like to express our sincere thanks to Dr. Fook-Long Lo (Universiti Tunku Abdul Rahman, Malaysia) for spending many hours polishing the manuscript.

E. H. LIM
K. W. LEUNG

City University of Hong Kong
Kowloon, Hong Kong SAR
January 8, 2012

CHAPTER ONE

Compact Multifunctional Antennas in Microwave Wireless Systems

1.1 INTRODUCTION

The mission of a communication system is to get messages delivered with minimum distortion. Messages such as voices, pictures, and movies are a series of natural signals over time, operating at frequencies ranging from a few to hundreds of kilohertz. Figure 1.1 shows the signal flows in a communication system. There are two types of communication systems: wired and wireless. Examples of wired systems are telephony and optical systems in which cables and fibers are deployed for transmitting signals, respectively. The telephone, patented by Alexander Graham Bell in 1876 [1], was the earliest available communication gadget that enabled the conversion of vocal messages into electronic signals. In 1966, Charles Kao [2] showed that a glass strand is able to be made into a signal-transmitting medium. Since then, tens of thousands of miles of optical fibers have been laid to carry information on land and across the oceans. The rapid advancement of optical technologies makes possible the transmission of signals in bulk using light, and it has led to a surge of internet technologies since the last century. However, the major drawback of wired communications is that it does not allow user mobility. Geographical features and human-made constructions can also pose a hindrance for laying out long wires or cables. As early as 1900, it was shown by Guglielmo Marconi that an electromagnetic wave is able to carry signals

Compact Multifunctional Antennas for Wireless Systems, First Edition. Eng Hock Lim, Kwok Wa Leung.
© 2012 John Wiley & Sons, Inc. Published 2012 by John Wiley & Sons, Inc.

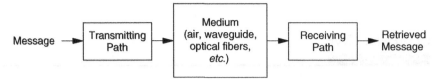

FIGURE 1.1   Signal flows in a communication system.

through air and free space. Since then, numerous analog and digital wireless communication systems have been developed. Figure 1.2 shows a typical analog wireless system, which has many functional blocks performing complex operations such as reception, transmission, modulation, and demodulation. As can be seen from the figure, the transmitting path consists basically of a modulator and a radio-frequency (RF) transmitter, while the receiving path has a demodulator and an RF receiver. In an analog wireless system all the signals are continuous. As shown in Fig. 1.3, the system can easily be made digital by incorporating analog-to-digital and digital-to-analog converters. In modern digital wireless systems, the modulation, demodulation, coding, and decoding processes can be performed easily by superfast microprocessors and digital signal processors. An advantage of digital signal is that many powerful coding schemes, such as the Viterbi, Trellis, and Turbo codes, can easily be imposed on the signal sequence (in "0" or "1") to enhance its robustness against noise [3]. The coding process is usually accomplished by connecting an encoder to the transmitting path and a decoder to the receiving path simultaneously. The encoder can be a circuit, a software program, or firmware (an algorithm burned into programmable hardware) that *converts* the source bits to channel bits. On the other end, a decoder is employed to retrieve the original message from the channel bits received. Various security features can also be added during the encoding–decoding process. As the encoder and decoder

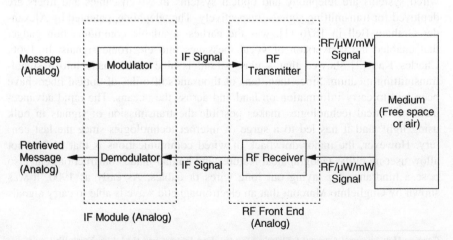

FIGURE 1.2   Typical analog wireless communication system.

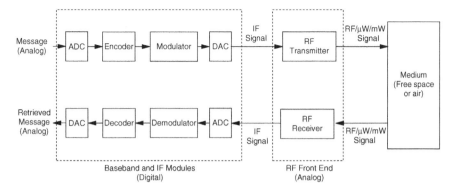

FIGURE 1.3 Typical digital wireless communication system.

do not change the fundamental frequencies of a message signal, they are usually called baseband modules.

Modulation is a process of transforming the spectrum of the baseband signal to a higher frequency, called the intermediate frequency (IF). It can be used to optimize bandwidth usage and enhance the signal quality during transmission. Before transmission, the signal is moved to an even higher frequency in the RF transmitter and then sent through the channel. At the receiver, after the step down from RF, a demodulator is used to retrieve the message from the IF signal received. Both the modulator and demodulator usually work at a frequency ranging from several kilohertz to hundreds of megahertz, which is the IF range. In this frequency range, circuits and systems can be designed simply using lumped components without involving transmission-line techniques. It can be seen from Figs. 1.2 and 1.3 that modulation and demodulation can be performed either in analog or digital form. Some of the famous analog modulation schemes are amplitude modulation (AM), frequency modulation (FM), and phase modulation (PM). Today, these schemes are still being used by many commercial radio stations. As can be seen from Fig. 1.3, the digital modulator and demodulator are used in the baseband and IF modules of a digital wireless system. Digital modulation schemes such as ASK (amplitude shift keying), PSK (phase shift keying), FSK (frequency shift keying), GMSK (Gaussian minimum shift keying), and OFDM (orthogonal frequency-division multiplexing) are among the popular choices in contemporary digital wireless systems.

With reference to Figs. 1.2 and 1.3, for both the analog and digital wireless communication systems, the output signals of the transmitters are always continuous with frequencies in the RF, microwave (μW), or millimeter-wave (mW) ranges. This is because antennas can be used to convert the signals in these frequency ranges into electromagnetic (EM) waves for propagation in air, which is a common channel medium for wireless communications. After traveling for a long distance in the channel (either air or free space), an EM wave arrives at the receiving antenna of an RF receiver. The weak and noisy signal received is finally demodulated and decoded so that the original message signal can be retrieved. The RF transmitter and receiver are generally called the RF front end, as they work at the RF/μW/mW frequency ranges, starting from several hundred megahertz up to tens of gigahertz. Since there are many wireless signals in air, proper allocation of the frequency spectrum is needed to avoid any chaos. To this end, wireless communication protocols such as BT (Bluetooth), DECT (digital enhanced cordless communication telecommunication), GSM (global system for mobile communication), GPRS (global packet radio service), IMT-A (international mobile telecommunications–advanced), UMTS (universal mobile telecommunications system), WiBro (wireless broadband), WiMax (worldwide interoperability for microwave access), and WLAN (wireless local area network) use different parts of the frequency spectrum. The spectrum allocation charts for some commercial mobile and satellite communication systems are given in Tables 1.1 and 1.2, respectively. The same spectrum can also be used simultaneously by many users by applying additional schemes, such as TDMA (time-division multiple access) and CDMA (code-division multiple access).

In this book we discuss only RF transmitters and receivers. The architecture of a typical one-stage unilateral RF transmitter is shown in Fig. 1.4(a). By incorporating a local oscillator (LO), the UP mixer can scale up the frequency of an IF signal. The role of the local oscillator is to impose an RF signal, usually called a carrier, onto the IF signal. Then a power amplifier is deployed for boosting the signal strength for transmission over a greater distance. With reference to Fig. 1.4(a), bandpass (image) filter has been used to remove the unwanted image signals generated by

TABLE 1.1   Frequency Bands (MHz) Allocated for Some Popular Mobile Communication Systems

Modulation Scheme	Uplink	Downlink
GSM-850	824–849	869–894
GSM-900	890–915	935–960
GSM-1800	1710–1785	1805–1880
GSM-1900	1850–1910	1930–1990
UMTS	1885–2025	2110–2200
WiMax	2300–2500 and 3400–3500	
WiBro	2300–2400	
WLAN, Bluetooth	2400–2480	

TABLE 1.2   Frequency Bands (GHz) Allocated for Satellite Communications

Band	Uplink	Downlink	Users
C	5.925–4.425	3.7–4.2	Commercial
X	7.9–8.4	7.9–8.4	Military
Ku	14.0–14.5	11.7–21.2	Commercial
Ka	27.5–30.5	17.7–21.2	Military
Q	43.5–45.5	20.2–21.32	Military

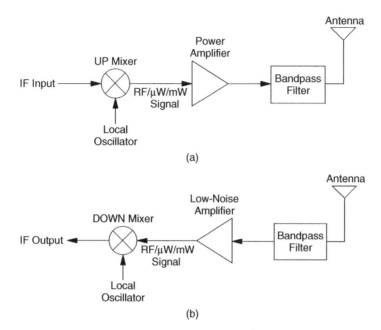

FIGURE 1.4   (a) Unilateral RF transmitter; (b) unilateral RF receiver.

the UP mixer. Finally, through the use of a transmitting antenna, the RF signal is channeled into the air. At the unilateral RF receiver shown in Fig. 1.4(b), bandpass filter is used to remove the unwanted signals and noise picked up by the RF signal from the channel medium. A low-noise amplifier (LNA) is then inserted to magnify the signal received, which is usually weak and noisy after traveling a long distance in the channel. Finally, a local carrier signal is used to down-convert the RF signal back to IF so that it can be processed by other modules. Multiple stages can easily be cascaded to achieve better performances. For a modern wireless system, the RF front ends are required to be low loss, low cost, light weight, high performance, power efficient, and small in size.

In modern wireless communication systems, the RF transmitter and receiver are often combined with a modulator and demodulator to form a single-module transceiver. The architecture of a typical bilateral transceiver [4] is shown in Fig. 1.5. Except for the antenna, all the components in a transceiver can be made easily on a single silicon chip. As a result, the antenna is usually the bulkiest component in a transceiver. It is always very desirable to have as few antennas as possible in a wireless communication system. With reference to Fig. 1.5, a switch has been employed so that an antenna can be shared by the transmitter and receiver. The preset filter here is a bandpass filter for removing channel noise. With reference to the figure, the RF modulator and demodulator here are implemented by incorporating an UP/DOWN mixer with a voltage-controlled oscillator (VCO) and a synthesizer. The two bandpass filters are to remove the unwanted frequency

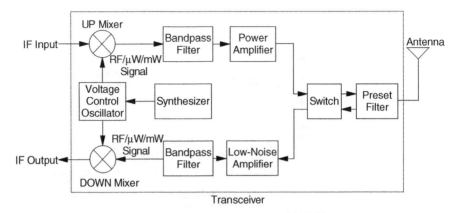

FIGURE 1.5   Bilateral transceiver structure. [4]

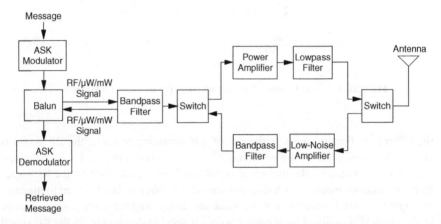

FIGURE 1.6   Block diagram of a commercial RF transceiver module. [5]

components. Figure 1.6 shows the functional blocks of a contemporary commercial RF transceiver [5], which is implemented using the ASK modulation scheme. The architecture is very close to that shown in Fig. 1.5. Despite the rapid advancement of semiconductor technologies, for some special applications, many components in the RF front end still remain discrete in order to achieve a high quality factor [6]. Usually, microwave components such as antennas, filters, and couplers are among the bulky components which are difficult to miniaturize, as they may involve the use of various microwave resonators.

1.2   MICROWAVE COMPONENTS IN WIRELESS SYSTEMS

There are two types of microwave components in the RF front end of a wireless system: active and passive. Active elements such as amplifiers, switches, oscillators,

TABLE 1.3   Features of the Microwave Components That Are Made by ICs and EM Structures

	Integrated Circuits	EM Structures
Circuit size	Very small (<1 mm^2) [8,9]	Large ($\sim$cm^2, depending on operating frequency)
Q factor	Low (<10) [10]	Very high (>100)
Costs	More expensive	Less expensive
Fabrication process	Complex	Easy

and mixers, made by employing semiconductor integrated circuit (IC) technologies, are necessary for performing power amplification and conversion. Nonlinear devices such as bipolar-junction transistors (BJTs), complementary metal–oxide semiconductors (CMOSs), and other transistors are commonly used to build these active circuits on various semiconductor materials, such as silicon, germanium, or gallium arsenide (GaAs). Nowadays, passive components such as capacitors and inductors can be easily fabricated on semiconductor substrate. For example, a capacitor can be made very compact by making use of the gate–substrate capacitance of a transistor. However, an inductor always occupies a large chip area, as it requires metallic coils to produce sufficient magnetic field [7].

Recent rapid advancement of IC fabrication processes has enabled significant size miniaturization in the RF front end. Baseband and IF modules such as encoders, decoders, modulators, and demodulators, are usually fabricated on ICs. Nevertheless, passive elements such as filters, baluns, circulators, and antennas are still difficult to miniaturize, as they are usually made on various wave-guiding resonators. Although some of these microwave components may possibly be designed by cascading lumped capacitors and inductors on a chip, it is still not easy to obtain a high Q factor, due to the lossy nature of semiconductor. As a result, they are generally made on microwave resonators for squeezing up the Q factor, which is essential for improving the frequency selectivity of the RF front end. A simple comparison of the microwave components that are made by using ICs and EM structures is shown in Table 1.3.

1.3   PLANAR AND NONPLANAR ANTENNAS IN COMPACT WIRELESS SYSTEMS

An antenna is the only component in the RF front end that cannot be made simply by using lumped components, as it requires a certain physical mechanism to enable efficient electromagnetic radiation. It is well known that acceleration and deceleration in electron flows are essential for generating EM waves [11]. There are many types of antennas. The most common antennas are electromagnetic radiators, which can be made on either a resonator or a traveling-wave structure. A resonating antenna has standing waves formed in its resonator. Some simple examples of resonating antennas are the monopole, dipole, aperture, and microstrip patch. On the other hand, traveling-wave antennas make use of the nonresonant voltages and currents on transmission lines to produce EM radiation. Examples of such antennas are the Beverage, Yagi–Uda, log-periodic, helix, dielectric rod, rhombic, spiral, and horn antennas. Traveling-wave antennas can also be classified as either slow-wave (surface-wave antennas) or fast-wave (leaky-wave antennas) antennas, according to the wave velocity.

1.3.1 Performance Parameters

A transmitting antenna is a device that converts alternating current (ac) into an EM wave for propagation in space. Therefore, a good antenna is not only required to have good circuit performance but must also satisfy many stringent far-field requirements [11,12]. Here, we discuss briefly some of the important design parameters for an antenna.

Antenna Bandwidth   Figure 1.7 shows the equivalent circuit of an antenna (represented by its input impedance Z_{ant}) and its interconnecting cable (with a characteristic impedance Z_0 and propagation constant β) with a length of l. P'_{in} is the total input power, P_{in} is the power being transferred to the antenna, and P_r is the radiated power. The reflection coefficient at the antenna port is defined as

$$S_{11} = \frac{Z_{ant} - Z_0}{Z_{ant} - Z_0} \tag{1.1}$$

which can easily be measured using a vector network analyzer. The power being delivered to the antenna is

$$P_{in} = (1 - |S_{11}|^2)P'_{in} \tag{1.2}$$

With reference to Fig. 1.8, the antenna bandwidth is usually defined as the frequency range where $|S_{11}| \leq -10$ dB (VSWR = 2). This is also considered to be

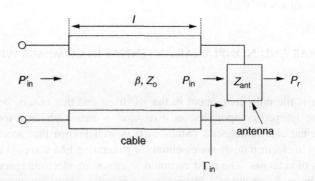

FIGURE 1.7   Equivalent circuit of an antenna connected to a cable.

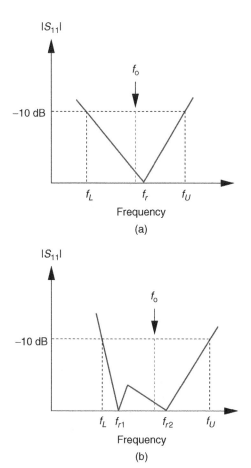

FIGURE 1.8   Definition of the antenna bandwidth for a radiator with (a) a single mode; (b) multiple modes.

the operating frequency range of an antenna. As can be seen from the figure, f_r (or $f_{r1}, f_{r2}, \ldots, f_{rN}$ for a multimode antenna) is the resonance frequency, f_U is the upper bound, and f_L is the lower bound of an antenna passband. For a single-mode antenna with $\Delta f < 100\%$, shown in Fig. 1.8(a), either (1.3) or (1.4) can be used to determine the antenna bandwidth Δf, where the center frequency is defined as $f_0 = (f_U + f_L)/2$. However, (1.3) is not suitable for a multimode antenna, which may have many resonant frequencies. The definition stated in (1.5) is another common definition used in describing the antenna bandwidth of an ultrawideband (UWB) antenna. For example, a wideband antenna covering 3.1 to 10.3 GHz has a bandwidth of 3.3:1. It is always desirable to have a wide antenna bandwidth.

$$\Delta f = \frac{f_U - f_L}{f_r} \times 100\% \qquad (1.3)$$

$$\Delta f = \frac{f_U - f_L}{f_0} \times 100\% \tag{1.4}$$

$$\frac{f_U}{f_L} : 1 \tag{1.5}$$

The loss mechanism of an antenna is characterized by its unloaded quality factor (Q_u), which consists of radiation loss (Q_r), ohmic loss (Q_c), and dielectric loss (Q_d). It can be expressed as

$$\frac{1}{Q_u} = \frac{1}{Q_r} + \frac{1}{Q_c} + \frac{1}{Q_d} \tag{1.6}$$

For an EM-wave radiating element, Q_r is the dominant factor for calculating Q_u. A lower Q_u implies a higher radiation power P_r. As an antenna is always connected to an external circuit, the loaded Q_L factor is used more frequently. As can be seen from (1.7), the Q_L can be calculated directly from f_r and Δf for a single-mode antenna:

$$Q_L = \frac{f_r}{\Delta f} \tag{1.7}$$

Antenna bandwidth (Δf) is inversely proportional to the minimum quality factor Q_{min} factor of an antenna. It is stated by Chu's criterion [13] that Q_{min} is related to the operating frequency (k_o) and the radius (R) of a virtual sphere that encloses the antenna by

$$Q_{min} = \frac{1 + 3(k_o R)^2}{(k_o R)^3 [1 + (k_o R)^2]} \tag{1.8}$$

Radiation Efficiency   The radiation efficiency (e) is a figure of merit showing how efficient energy is being converted from input power (P_{in}) to output radiation (P_r). P_L is the dissipative loss. With reference to Fig. 1.9, e can be described by

$$e = \frac{P_r}{P_{in}} \tag{1.9}$$

The total input power can be written as $P_{in} = P_r + P_L$. Assuming that the current flowing into the antenna is I_{in}, the radiated and dissipated powers are defined as $P_r = I_{in} R_r$ and $P_L = I_{in} R_L$, respectively, where R_r (radiation resistance) and R_L (loss resistance) are the virtual resistors representing the radiated and dissipated powers. The values are then substituted into (1.9) to obtain

$$e = \frac{P_r}{P_r + P_L} = \frac{I_{in} R_r}{I_{in}(R_r + R_L)} = \frac{R_r}{R_r + R_L} \tag{1.10}$$

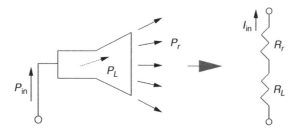

FIGURE 1.9   Radiated and dissipated powers of an antenna.

Radiation efficiency includes both conduction and dielectric losses, which are usually not directly measurable. There are a number of methods for measuring radiation efficiency, the simplest and most widely used technique being that proposed by Wheeler in 1959, called the Wheeler cap method [14,15]. It is suitable for both single- and multimode resonating antennas [16,17]. The radiation efficiency of a small antenna can be estimated by calculating the power dissipated by an antenna [18,19]. The antenna temperature has also been used to measure the radiation efficiency [20,21]. It was recently shown that this parameter is also obtainable by discerning the time-domain response of an antenna to a short pulse [22]. Various measurement methods have been compared by several authors [23,24].

Radiation Pattern   The radiation pattern describes a collective of electric (or magnetic) field strengths at a fixed distance in the far field of an antenna. Figure 1.10 shows the electric far fields radiated by an illustrative antenna placed at the origin of a spherical coordinate system $(\vec{a}_r, \vec{a}_\theta, \vec{a}_\phi)$. Since the antenna dimension is usually much smaller than the far-field distance (R), it can conveniently be considered to be a point source. For a simple antenna, the far-field distance is usually defined

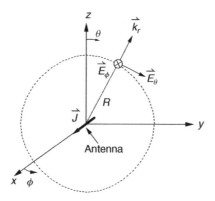

FIGURE 1.10   Antenna placed at the origin of a spherical coordinate system.

TABLE 1.4 Major Cut Planes in the Radiation Pattern of an Antenna

(x, y, z)	$(r = $ Far Field, ϕ, $\theta)$	Co-polarized Field, $\vec{E}_{co}$	Cross-Polarized Field, $\vec{E}_{cross}$
xz-plane (E-plane)	$\phi = 0°$	$\vec{E}_\theta$	$\vec{E}_\phi$
yz-plane (H-plane)	$\phi = 90°$	$\vec{E}_\phi$	$\vec{E}_\theta$

as $R \geq 2D^2/\lambda_0$, where D is the largest dimension of an antenna and λ_0 is the operating wavelength in air.

According to the Poynting vector, only $\vec{E}_\theta$ and $\vec{E}_\phi$ components contribute to wave propagation in the radial direction ($\vec{k}_r$). For a symmetrical radiation pattern, the field can usually be described sufficiently by its components on the three major cut planes (xy, xz, and yz). The plane that is parallel to the vector of the maximum electric field is defined as the E-plane. Similarly, the cut plane that contains the maximum magnetic vector is called the H-plane. Consider an antenna that is equivalent to a simple electric current ($\vec{J}$) flowing in the x-direction, Table 1.4 shows the coordinate systems of its major cut planes, along with the corresponding co- and cross-polarized fields.

For an antenna that has a symmetric radiation pattern with respect to the z-axis, Fig. 1.11 depicts its radiation patterns at an arbitrary cut plane $\phi = \phi'$. In the E-plane ($\phi' = 0°$), the field components $\vec{E}_\theta$ and $\vec{E}_\phi$ are the co- and cross-polarized fields, respectively. The same definition applies to the H-plane ($\phi' = 90°$), with $\vec{E}_\phi$ defined as the co-polarized field and $\vec{E}_\theta$ as the cross-polarized field. The definitions of some of the key parameters for a radiation pattern are now discussed. With reference to Fig. 1.11, the *half-power beamwidth* is the angular range in which $|\vec{E}_{co}|^2 \geq 0.5|\vec{E}_{max}|^2$. It is referred to simply as the 3-dB beamwidth. *Antenna gain* is defined as the ratio of the co-polarized field to that of a hypothetical isotropic radiator. The ratio between the co- and cross-polarized fields in a particular angular direction is called the *cross-pole rejection*. Higher rejection implies purer field polarization. For a broadside antenna with the radiation pattern shown in Fig. 1.11, the *front-to-back* (FB) *ratio* is defined as $\vec{E}_{co}(\theta = 0°)/\vec{E}_{co}(\theta = 180°)$, where $\vec{E}_{co}(\theta = 0°)$ is the co-polarized field in the boresight direction and $\vec{E}_{co}(\theta = 180°)$ is that for the back side. A high FB ratio is very desirable for a high-gain antenna. All the aforementioned parameters vary with frequency. Directional (broadside) and omnidirectional (end-fire) antennas are among the most popular EM radiators. Directional antennas are deployed for communication systems where a distant point-to-point connectivity is desired. On the other hand, an omnidirectional antenna is used for communication in multiple directions, such as that for a hand phone.

Polarization Antennas can generally be categorized as linear, circular, or elliptical polarized antennas, based on the motion of the propagating electric field with respect to time and space. Polarization matching between the transmitter and the receiver is crucial for good signal reception. With reference to Fig. 1.12(a)

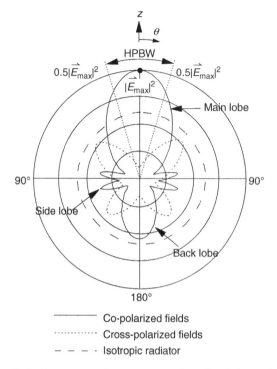

FIGURE 1.11   Radiation pattern of an antenna at the azimuthal cut plane of $\phi = \phi'$.

and 1.12(b), the plane wave propagating in the z-direction is called a *linearly polarized wave*, as its electric field only sweeps at all times in the x (or y)-direction [12]. A circular polarized wave propagates in the z-direction with its electric field ($\vec{E}$) rotating circularly around the z-axis, as can be seen in Fig. 1.12(c). In this case, $\vec{E}$ can be decomposed into two equal components, $\vec{E}_1$ and $\vec{E}_2$. The ratio $\vec{E}_2/\vec{E}_1$ is defined as the *axial ratio* (AR). A perfect circular polarization $\vec{E}_1 = \vec{E}_2$ implies that AR = 1 (or 0 dB) and the 3-dB AR bandwidth is defined as the frequency range where AR $\leq$ 3 dB. When $\vec{E}_1 \neq \vec{E}_2$, the trajectory of $\vec{E}$ is an ellipse, as shown in Fig. 1.12(d). For both the circular and elliptical cases, an electric field in clockwise motion is referred to as *left-handed polarization* and that in anticlockwise motion as *right-handed polarization*.

An antenna is not a stand-alone device in wireless systems. The design of an antenna is always coupled with the requirements of the entire system. It is always desirable to have an antenna that is compact, aesthetic, low profile, and light weight. Being low cost and easy to design and tune are also important design considerations that have to be taken into account. As an antenna radiates, the operation of other devices may get affected. Therefore, a good antenna must be compatible with other electronic components; most important, it is not expected to disrupt the normal operation of other systems. Mobile gadgets such as hand phones and walkie-talkies are usually placed close to the human body. In these applications, the

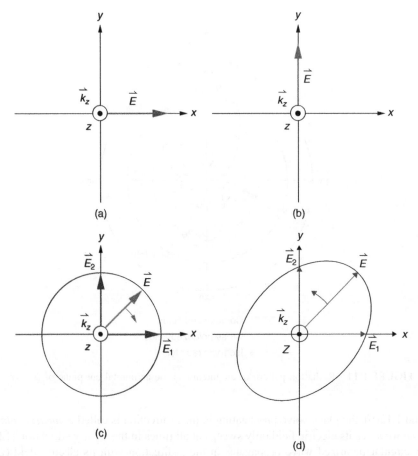

FIGURE 1.12   (a) Linear polarization in the x-direction; (b) linear polarization in the y-direction; (c) left-handed (clockwise) circular polarization; (d) right-handed (anticlockwise) elliptical polarization.

antenna has to be designed carefully so that the body is not exposed to excessive EM radiation. The Federal Communications Commission (FCC) requires that all phones sold in the United States have an SAR (special absorption rate) $\leq$ 1.6 W/kg for 1 g of human tissue. In Europe, the CENELEC (European Committee for Electrotechnical Standardization) sets the SAR limit as 2 W/kg averaged over 10 g of tissue. In Table 1.5, the design specifications are given for a commercial microstrip antenna. Circular polarization is commonly used by global positioning systems (GPSs).

1.3.2   Planar Antennas

Planar antennas are EM radiators that are flat in shape or, at least, conform to a curved surface. This feature enables easy integration with other surface-mounted

TABLE 1.5 Design Specifications for a Commercial Microstrip Patch Antenna for GPS Application

	Minimum	Typical	Maximum	Unit
Frequency	1570.42	1575.42	1580.42	MHz
Polarization	—	Right-handed	—	RHCP
Radiation efficiency	—	65	—	%
Maximum antenna gain	—	4.8	—	dBi
HPBW	—	90	—	degrees
Cross-pole rejection	—	19	—	dB
Axial ratio	—	—	3	dB
FB ratio	—	25	—	dB
VSWR	—	1.3:1		
Impedance (differential)	—	50	—	Ω
Operating temperature	−40	20	+85	°C
Overall dimensions		$12 \times 12 \times 4$		mm

Source: [25].

components. Microstrip patch antennas, slot antennas, planar inverted-F antennas (PIFA), planar inverted-L antennas (PILA), chip dipoles and monopoles, and suspended plate antennas (SPAs) are among the popular antennas used extensively by wireless communication systems [13,26,27]. Figure 1.13 shows some planar antennas.

(a)

FIGURE 1.13 Examples of planar antennas: (a) planar dipole array; (b) circular patch; (c) microstrip patch array; (d) RFID.

1.3.3 Nonplanar Antennas

Figure 1.14 shows some of common nonplanar antennas, which are usually three-dimensional in structure [28]. Because of their bulkiness, such antennas are usually installed external to the circuit part. In many cases, nonplanar antennas require the use of reflectors, directors, impedance transformers, or multiple elements for better antenna performance. Examples of nonplanar antennas are disk antennas, horns, helixes, and reflector antennas.

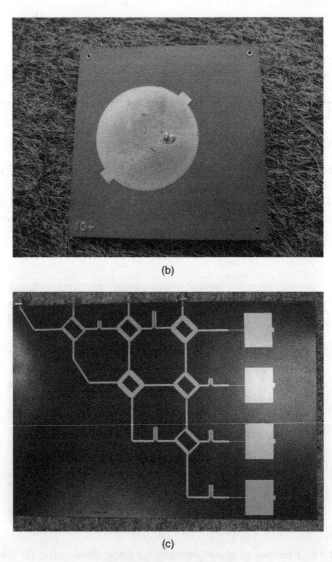

(b)

(c)

FIGURE 1.13 (*Continued*)

(d)

FIGURE 1.13 (*Continued*)

1.4 MULTIFUNCTIONAL ANTENNAS AND MICROWAVE CIRCUITS

A multifunctional antenna is an EM radiating element that provides additional microwave functions. It is also known as an *antenna-circuit module*. Similarly, a multifunctional microwave circuit is one that has multiple functions in one module and is also called a *circuit–circuit module*.

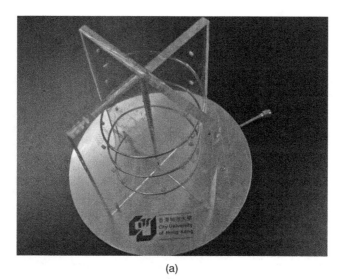

(a)

FIGURE 1.14 Examples of nonplanar antennas: (a) helix antenna; (b) spiral antenna; (c) loop antenna; (d) monopolar-type antenna.

1.4.1 Active Antennas

An active integrated antenna (AIA), being multifunctional, is an EM radiating element that provides at least one built-in function, such as amplification, equalization, oscillation, mixing, modulation–demodulation, reconfiguration, and switching [29,30]. These additional features are provided by incorporating an antenna with active devices such as diodes, switches, and various transistors. Usually, AIAs are designed without the use of matching circuits, and therefore they can be made very compact. Active antennas have been explored extensively since the 1980s for

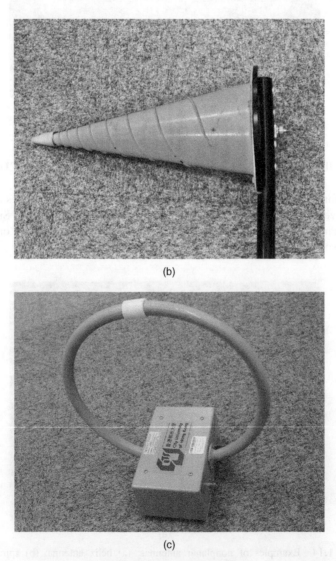

(b)

(c)

FIGURE 1.14   (*Continued*)

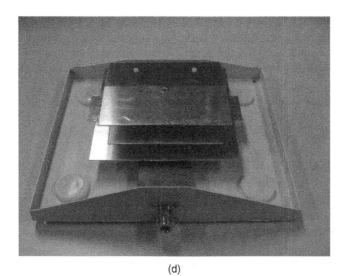

(d)

FIGURE 1.14   (*Continued*)

applications in spatial power combining, beam steering, and retro-directive arrays [29,31,32].

1.4.2 Passive Antennas

Passive integrated antennas and passive integrated circuits are very popular because of a number of advantages, such as their low loss, low cost, and high compactness. Nowadays, it is a trend to combine several microwave functions into a single module for a compact footprint [33,34]. In recent years, new components have been proposed, including the antenna filter [35], antenna circulator, antenna package [36], and balun filter [37]. With the rapid advancement of packaging technologies, new design concepts, such as antenna-on-package (AOP), system-on-package (SOP), antenna-in-package (AiP), system-in-package (SiP), and low-temperature co-fired (LTCC) have been put forward. These are discussed in detail in Chapter 2.

1.5 MINIATURIZATION TECHNIQUES FOR MULTIFUNCTIONAL ANTENNAS

Bundling several microwave functions into a single module is definitely a good way to reduce circuit size and footprint. The following techniques have frequently been used to miniaturize multifunctional antennas and microwave circuits:

1. *Sharing a single resonator*. Different resonances can be excited in a single resonator for designing different microwave components. Sometimes, a resonator can be used simultaneously for several functions. In general, if two microwave functions can be combined, it simply cuts down the circuit size by half. Jung and Hwang [34] made a balun and a filter on a ring resonator. In [35], the $TE_{01\delta}$ and $HEM_{11\delta}$ modes of a single dielectric resonator (DR) are used simultaneously for a dielectric resonator antenna and a dielectric resonator filter. It is also shown that a hollow DR can be used as a packaging cover at the same time [36].

2. *Combining multiple components*. Combining the resonators of several microwave devices into a single platform is a common technique used to miniaturize circuits. The devices are usually squeezed into the two surfaces of a substrate. In [37], a microstrip patch antenna and an open-ring filter are combined so that they can share the same footprint. It was shown that a balun can be placed on the substrate underside of a printed dipole [38] to form a compact RF module.

3. *Applying multilayer technology*. Modern packaging techniques enable easy fabrication of various multilayer configurations. AOP, AiP, SOP, SiP, and LTCC are among the popular technologies used for packing multiple microwave components into multilayer packages [39,40]. In principle, the entire RF front end can be made into different layers of a single package. For these configurations, antennas are usually stacked on top of the packages to enable EM radiation.

4. *Integrating into substrate*. Recently, substrate integration technology has been widely used in the design of both active and passive multifunctional modules in the millimeter-wave range. In [41], a substrate integrated waveguide (SIW) cavity-backed slot antenna oscillator is proposed. Using a half-mode SIW, Cheng et al. [42] proposed a quadri-polarized frequency scanning antenna built by combining a 3-dB coupler and a leaky-wave antenna array.

5. *Removing matching circuits*. This technique is used frequently by amplifying antennas [43]. On the receiver side, the active low-noise amplifying antenna usually does not contain any input-matching circuit. On the transmitter side, a power-amplifying antenna works without having an output-matching circuit [44]. Removal of part of the peripheral circuit helps to cut down the component size and signal path.

6. *Multitasking the microwave element*. For the design of antenna oscillators, the antenna can be used simultaneously as the resonator, load [45], and feedback element [46] of an oscillator. Squeezing multiple tasks into a microwave element can reduce the circuit size significantly.

1.6 DESIGN PROCESSES AND CONSIDERATIONS

Some topics pertaining to antenna–circuit modules have been covered by Gupta and Hall [47]. Recently, the rapid advancement in computer-aided design (CAD) tools, along with the availability of more powerful computers, has significantly shortened the design cycle of multifunctional antennas and microwave circuits [48]. Designing a multifunctional component usually involves co-designing and co-optimizing several microwave functions. Flow diagrams depicting the design processes are shown in Fig. 1.15. CAD tools can be used in many ways to assist in the design process. For an antenna–circuit module, the antenna and circuit are usually designed separately, as they may require different modeling approaches. The calculation of antenna far fields requires full-wave analysis, such as the moment method (MoM), the finite integral technique (FIT), the finite-element method (FEM), the time-domain finite-difference method (FDTD), geometric optics (GO), physical optics (PO), and the uniform theory of diffraction (UTD). On the other hand, circuit- and network-based methods, such as nodal analysis, the transmission matrix, and harmonics balance, are frequently used to analyze both active and passive circuits. Over the past two decades, much effort has been devoted to developing various optimization algorithms for microwave applications. The genetic algorithm [49,50], particle swarm [51,52], and neural network [53] are

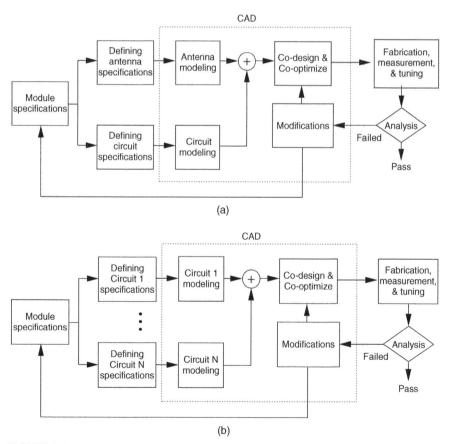

FIGURE 1.15 Design procedures of multifunctional integrated circuits: (a) antenna–circuit module; (b) circuit–circuit module.

among the popular optimization algorithms frequently incorporated with various computational EM tools to expedit the optimization process. Recently, stochastic methods [54,55] have also been explored for optimizing complex EM problems.

Despite the availability of many powerful CAD tools, combining antennas with circuits still requires special care, as the joint simulation model can become very large and complex. In computer modeling, there is always a trade-off between completeness and accuracy. A complete full-wave model is good for accuracy, but it requires many more computational resources and a longer simulation time. In contrast, a simplified model is easily manageable by computers, but at the expense of accuracy. To expedite the simulation time, it is also not uncommon to omit the coupling effects of antenna-to-antenna and antenna-to-circuit connections. This can again introduce errors in the computation.

1.7   DESIGN TOOLS AND SOFTWARE

Today, many CAD softwares, usually adorned with colorful and user-friendly graphic user interfaces, are commercially available for modern RF and microwave designs. They are commonly used to reduce the design-to-product time. In general, there are two types of microwave CAD tools: network (or circuit)- and field-based. An early software program was SPICE (simulation program with integrated-circuit emphasis). In the 1970s, computerized design tools were less common, and most of microwave design programs were encoded in computer programs or punch cards, often owned by large IC manufacturers and customized only for certain transistors. For example, SPEEDY was developed by Les Besser for Fairchild, and CAIN-01 was written by R. P. Coats for Texas Instruments. In 1973, Besser introduced the first commercially available CAD software, called COMPACT (computer optimization of microwave passive and active circuits), for designs of various microwave circuits [56]. It was later converted into SuperCOMPACT and became an industrial standard at that time.

Since then, a 40-year period of computer development has brought CAD technology to another height. Some commonly used commercial EM software is compared in Table 1.6. Today, many complex linear and nonlinear circuits, such as filters, couplers, amplifiers, oscillators, and mixers, can easily be modeled using two-dimensional SPICE software, which is supported by a vast number of device libraries. As can be seen in the table, many three-dimensional full-wave tools are also available for simulating field distributions of antennas and EM structures.

The co-design and co-optimization processes may require the use of multiple CAD tools. The system outputs of a microwave structure (either a passive circuit or an antenna), which are usually simulated using a three-dimensional full-wave software, can usually be exported to circuit- and network-based CAD tools for co-simulations. As an example, it was shown by Lim et al. [33] that the reflection coefficient S_{11} (or input impedance Z_{in}) of an antenna can be imported directly into the AWR Microwave Office as a subcircuit module for combination with other devices. Today, many advanced three-dimensional full-wave EM softwares, such as

TABLE 1.6 Comparison of Some Commercially Available RF and Microwave Design Softwares

Software	Company	Method	Applications
Ansoft HFSS [57]	Ansys	Frequency-domain FEM	Three-dimensional full-wave EM simulators for both planar and nonplanar antennas and passive circuits
CST Microwave Studio [58]	Computer Simulation Technology	Time- and frequency-domain FIT	
FEKO [59]	EM Software & Systems - S.A. (Pty) Ltd	Hybridized with MoM, FEM, PO, GO, and UTD	
IE3D SSD [60]	Mentor Graphics	Frequency-domain MoM	
Sonnet [61]	Sonnet Software	MoM	
XFdtd [62]	REMCOM	Time-domain FDTD	
Momentum [63]	Agilent EEs of EDA	Frequency-domain MoM	Three-dimensional full-wave planar EM simulator suitable for passive circuit modeling and analysis
Advanced Design System [64]	Agilent EEs of EDA	Nodal-, circuit-, and network-based methods	Design of two-dimensional active—passive linear and nonlinear RF and microwave circuits and networks
Microwave Office [65]	AWR Corporation		

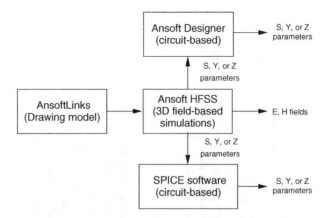

FIGURE 1.16   Simulation platform for Ansoft EM simulators. [57]

Ansoft HFSS and CST Microwave Studio, have their own built-in circuit-based simulators. Figure 1.16 shows the co-simulation platform for a series of Ansoft EM software [57]. As can be seen, the outputs (S, Y, or Z parameters) of an antenna (Ansoft HFSS) can easily be exported internally to the Ansoft Designer (circuit-based) or externally to another SPICE simulator for joint simulations. This simplifies the design processes of multifunctional antennas and circuits significantly.

1.8   OVERVIEW OF THE BOOK

This book encompasses multifunctional antennas and microwave components, along with their design methods. Recent developments and important design issues are also discussed. In this book, multiple design tools and softwares are used jointly for the design of multifunctional components which may contain both active and passive devices. To facilitate the reading, a brief overview of this book is provided here.

Chapter 2 presents several new multifunctional passive integrated antennas and passive integrated circuits, such as an antenna filter, antenna package, and balun filter. Dielectric resonators and microstrip-based resonators are used to demonstrate the design ideas. In most cases, multiple functions are implemented on a piece of single resonator for a compact footprint. With the use of field-based and SPICE softwares, design procedures for several multifunctional modules are described in detail.

In Chapter 3 it is shown that antennas can be made reconfigurable by integrating switches into the radiation aperture. Frequency, pattern, and multiple reconfigurations are implemented using patch and fractal antennas. A pattern reconfigurable leaky wave antenna is also shown. Some recent work conducted at the Institute of Applied Physics, University of Electronic Science and Technology of China, Chengdu, is described.

Chapter 4 explores low-noise amplifying antennas. In the first part, a wearable active antenna made on a textile is described. It is shown that low-noise amplifiers can be designed without using an input matching circuit. Also, matching techniques are included.

Reflection-amplifier and coupled-load antenna oscillators are visited in Chapter 5. A number of design examples are given along with their design procedures. It is shown that reasonable phase noise and high output power are achievable in both types of antenna oscillators.

In Chapter 6, the recent development of solar-cell-integrated antennas is reviewed. Several new antenna configurations are covered. Dielectric resonator antennas and suspended plate antennas are examined to demonstrate the design ideas. Solar cells can be used in the future to provide self-sustaining power to wireless communication systems.

REFERENCES

[1] A. G. Bell, "Improvement in telephony," U.S. patent 174,465, Mar. 7, 1786.

[2] K. C. Kao and G. A. Hockham, Dielectric-fibre surface waveguides for optical frequencies, *IEE Proc.*, vol. 133, no. 3, pp. 191–198, June 1986. (Originally published in the same journal in 1966.)

[3] http://en.wikipedia.org/wiki/Encoder.

[4] K. Feher, *Wireless Digital Communications: Modulation and Spread Spectrum Applications*. Upper Saddle River, NJ: Prentice Hall, 1995.

[5] Datasheet: RFW3M-PA Transceiver Module, *Vishay RF Waves*, Feb. 2009.

[6] M. Steer, *Microwave and RF Design: A Systems Approach*. Raleigh, NC: SciTech, 2009.

[7] H. Moon, J. Han, S. Choi, D. Keum, and B. Park, An area-efficient 0.13-μm CMOS multiband WCDMA/HSDPA receiver, *IEEE Trans. Microwave Theory Tech.*, vol. 58, pt. 2, pp. 1447–1455, May 2010.

[8] S. S. K. Ho and C. E. Saavedra, "A CMOS broadband low-noise mixer with noise cancellation," *IEEE Trans. Microwave Theory Tech.*, vol. 58, pp. 1126–1132, May 2010.

[9] P. Mak and R. P. Martins, "A 2×VDD-enabled mobile-TV RF front-end with TV-GSM interoperability in 1-V 90-nm CMOS," *IEEE Trans. Microwave Theory Tech.*, vol. 58, pp. 1664–1676, July 2010.

[10] C. Lee, T. Chen, J. D. Deng, and C. Kao, "A simple systematic spiral inductor design with perfected Q improvement for CMOS RFIC application," *IEEE Trans. Microwave Theory Tech.*, vol. 53, pp. 523–528, Feb. 2005.

[11] C. A. Balanis, *Antenna Theory Analysis and Design*, 2nd ed. New York: Wiley, 1997.

[12] J. D. Kraus and R. J. Marhefka, *Antennas for All Applications*, 3rd ed. New York: McGraw-Hill, 2003.

[13] Z. N. Chen and M. Y. W. Chia, *Broadband Planar Antennas: Design and Applications*. Chichester, UK: Wiley, 2006.

[14] H. Wheeler, "The radiansphere around a small antenna," *Proc. IRE*, vol. 47, no. 8, pp. 1325–1331, Aug. 1959.

[15] G. Smith, "An analysis of the Wheeler method for measuring the radiating efficiency of antennas," *IEEE Trans. Antennas Propag.*, vol. 25, no. 4, pp. 552–556, July 1977.

[16] H. Choo, R. Rogers, and H. Ling, "On the Wheeler cap measurement of the efficiency of microstrip antennas," *IEEE Trans. Antennas Propag.*, vol. 53, pp. 2328–2332, July 2005.

[17] C. Cho, I. Park, and H. Choo, "A modified Wheeler cap method for efficiency measurements of probe-fed patch antennas with multiple resonances," *IEEE Trans. Antennas Propag.*, vol. 58, pp. 3074–3078, Sept. 2010.

[18] R. H. Johnston and J. G. McRory, "An improved small antenna radiation-efficiency measurement method," *IEEE Antennas Propag. Mag.*, vol. 40, pp. 40–48, Oct. 1998.

[19] W. L. Schroeder and D. Gapski, "Direct measurement of small antenna radiation efficiency by calorimetric method," *IEEE Trans. Antennas Propag.*, vol. 54, pp. 2646–2656, Sept. 2006.

[20] J. Ashkenazy, E. Levine, and D. Treves, "Radiometric measurement of antenna efficiency," *Electron. Lett.*, vol. 21, no. 3, pp. 111–112, Jan. 1985.

[21] N. J. McEwan, R. A. Abd-Alhameed, and N. Z. Abidin, "A modified radiometric method for measuring antenna radiation efficiency," *IEEE Trans. Antennas Propag.*, vol. 51, pp. 2099–2105, Aug. 2003.

[22] A. Khaleghi, "Time-domain measurement of antenna efficiency in reverberation chamber," *IEEE Trans. Antennas Propag.*, vol. 57, pp. 817–821, Mar. 2009.

[23] D. M. Pozar and B. Kaufman, "Comparison of three methods for the measurement of printed antenna efficiency," *IEEE Trans. Antennas Propag.*, vol. 36, pp. 136–139, Jan. 1988.

[24] E. H. Newman, P. Bohley, and C. H. Walker, "Two methods for the measurement of antenna efficiency," *IEEE Trans. Antennas Propag.*, vol. 23, pp. 457–461, July 1975.

[25] Datasheet: MPA1575D124 Compact Microstrip Patch Antenna GPS L1 Band, Maxtenna Antenna Innovations Company.

[26] K. L. Wong, *Compact and Broadband Microstrip Antennas*. Hoboken, NJ: Wiley, 2002.

[27] K. L. Wong, *Planar Antennas for Wireless Communications*. Hoboken, NJ: Wiley, 2003.

[28] K. L. Wong, *Design of Nonplanar Microstrip Antennas and Transmission Lines*. New York: Wiley, 1999.

[29] J. A. Navarro and K. Chang, *Integrated Active Antennas and Spatial Power Combining*. New York: Wiley, 1996.

[30] M. Steer and W. D. Palmer, Eds., *Multifunctional Adaptive Microwave Circuits and Systems*. Raleigh, NC: SciTech, 2009.

[31] A. Mortazawi, T. Itoh, and J. Harvey, *Active Antennas and Quasi-optical Array*. Piscataway, NJ: IEEE Press, 1999.

[32] R. A. York and Z. B. Popovic, *Active and Quasi-optical Arrays for Solid-State Power Combining*. New York: Wiley, 1997.

[33] E. H. Lim, K. W. Leung, and X. S. Fang, "The compact circularly-polarized hollow rectangular dielectric resonator antenna with an underlaid quadrature coupler," *IEEE Trans. Antennas Propag.*, vol. 59, pp. 288–293, Jan. 2011.

[34] M. Jung and H. Hwang, "A balun-BPF using a dual mode ring resonator," *IEEE Microwave Guided Wave Lett.*, vol. 7, pp. 652–654, Sept. 2007.

[35] E. H. Lim and K. W. Leung, "Use of the dielectric resonator antenna as a filter element," *IEEE Trans. Antennas Propag.*, vol. 56, pp. 5–10, Jan. 2008.

[36] E. H. Lim and K. W. Leung, "Novel application of the hollow dielectric antenna as a packaging cover," *IEEE Trans. Antennas Propag.*, vol. 54, pp. 484–487, Feb. 2006.

[37] Y. J. Sung, "Microstrip resonator doubling as a filter and as an antenna," *IEEE Antennas Wireless Propag. Lett.*, vol. 8, pp. 486–489, Mar. 2009.

[38] E. Avila-Navarro, J. Anton, J. M. Blanes, and C. Reig, "Broadband printed dipole with integrated via-hole balun for WIMAX applications," *Microwave Opt. Tech. Lett.*, vol. 53, no. 1, pp. 52–55, Jan. 2011.

[39] Y. P. Zhang and D. Liu, "Antenna-on-chip and antenna-in-package solutions to highly integrated millimeter-wave devices for wireless communications," *IEEE Trans. Antennas Propag.*, vol. 57, pp. 2830–2841, Oct. 2009.

[40] R. Li, G. DeJean, M. Maeng, K. Lim, S. Pinel, M. Tentzeris, and J. Laskar, "Design of compact stacked-patch antennas in LTCC multilayer packaging modules for wireless applications," *IEEE Trans. Adv. Packag.*, vol. 27, pp. 581–589, Nov. 2004.

[41] F. Giuppi, A. Georgiadis, A. Collado, M. Bozzi, and L. Perregrini, "Tunable SIW cavity backed active antenna oscillator," *Electron. Lett.*, vol. 46, no. 15, July 2010.

[42] Y. J. Cheng, W. Hong, and K. Wu, "Millimeter-wave half mode substrate integrated waveguide frequency scanning antenna with quadri-polarization," *IEEE Trans. Antennas Propag.*, vol. 58, pp. 1848–1855, June 2010.

[43] D. Segovia-Vargas, D. Castro-Galán, L. E. García-Mu noz, and V. González-Posadas, "Broadband active receiving patch with resistive equalization," *IEEE Trans. Microwave Theory Tech.*, vol. 56, pp. 56–64, Jan. 2008.

[44] G. A. Ellis and S. Liw, "Active planar inverted-F antennas for wireless applications," *IEEE Trans. Antennas Propag.*, vol. 51, pp. 2899–2906, Oct. 2003.

[45] E. H. Lim and K. W. Leung, "Novel utilization of the dielectric resonator antenna as an oscillator load," *IEEE Trans. Antennas Propag.*, vol. 55, pp. 2686–2691, Oct. 2007.

[46] K. Chang, K. A. Hummer, and G. K. Gopalakrishnan, "Active radiating element using FET source integrated with microstrip patch antenna," *Electron. Lett.*, vol. 24, pp. 1347–1348, Oct. 1988.

[47] K. C. Gupta and P. S. Hall, *Analysis and Design Integrated Circuit Antenna Modules*. Hoboken, NJ: Wiley, 2000.

[48] H. J. Visser, *Approximate Antenna Analysis for CAD*. Chichester, UK: Wiley, 2009.

[49] A. Hoorfar, "Evolutionary programming in electromagnetic optimization: a review," *IEEE Trans. Antennas Propag.*, vol. 55, pp. 523–537, Mar. 2007.

[50] J. J. Sun, D. S. Goshi, and T. Itoh, "Optimization and modeling of sparse conformal retrodirective array," *IEEE Trans. Antennas Propag.*, vol. 58, pp. 977–981, Mar. 2010.

[51] N. Jin and Y. Rahmat-Samii, "Parallel particle swarm optimization and finite-difference time-domain (PSO/FDTD) algorithm for multiband and wide-band patch antenna designs," *IEEE Trans. Antennas Propag.*, vol. 53, pp. 3459–3468, Nov. 2005.

[52] S. Genovesi, A. Monorchio, R. Mittra, and G. Manara, "A sub-boundary approach for enhanced particle swarm optimization and its application to the design of artificial magnetic conductors," *IEEE Trans. Antennas Propag.*, vol. 55, pp. 766–770, Mar. 2007.

[53] Y. Kim, S. Keely, J. Ghosh, and H. Ling, "Application of artificial neural networks to broadband antenna design based on a parametric frequency model," *IEEE Trans. Antennas Propag.*, vol. 55, pp. 669–673, Mar. 2007.

[54] Z. Li, Y. E. Erdemli, J. L. Volakis, and P. Y. Papalambros, "Design optimization of conformal antennas by integrating stochastic algorithms with the hybrid finite-element method," *IEEE Trans. Antennas Propag.*, vol. 50, pp. 676–684, May 2002.

[55] W. Weng, F. Yan, and A. Z. Elsherbeni, "Linear antenna array synthesis using Taguchi's method: a novel optimization technique in electromagnetics," *IEEE Trans. Antennas Propag.*, vol. 55, pp. 723–730, Mar. 2007.

[56] L. Besser, "A fast computer routine to design high frequency circuits," *IEEE ICC Conference*, San Francisco, June 1970.

[57] http://www.ansoft.com/products/hf/hfss/.

[58] http://www.cst.com.

[59] http://www.feko.info/.

[60] http://www.mentor.com/electromagnetic-simulation/.

[61] http://www.sonnetsoftware.com/.

[62] http://www.remcom.com/xf7.

[63] http://www.home.agilent.com/agilent/product.jspx?nid=-34333.804583.00&lc= eng&cc=HK.

[64] http://www.home.agilent.com/agilent/product.jspx?cc=HK&lc=eng&ckey=1297113 &nid=-34346.0.00&id=1297113.

[65] http://web.awrcorp.com/.

CHAPTER TWO

Multifunctional Passive Integrated Antennas and Components

2.1   DEVELOPMENT OF PASSIVE INTEGRATED ANTENNAS AND COMPONENTS

In the past few decades, many advances have been reported in the development of active integrated antennas (AIAs) [1–5], which were designed by incorporating various active devices, such as an amplifier, mixer, oscillator, duplexer, or rectifier, in an antenna. Being multifunctional, an AIA integrates various signal-processing functions into an antenna to enhance the antenna bandwidth, increase the effective length of a short antenna, reduce the coupling of an array, and improve the noise factor [1,3]. Most important, such integration can help in cutting down the signal transmission path and hence reduce the chance of picking up additional electronic noise. Recently, rapid advancement in millimeter-wave technologies has drawn significant attention to the research and development of passive integrated components (PICs). In modern microwave systems it is very desirable to integrate several functions into a single module to achieve high compactness, low loss, and low cost. Generally, there are two types of multifunctional PICs, one of which combines multiple resonators to obtain multifunctionality, while another achieves multiple functions in a single resonator. In the past two decades, many multifunctional PICs, such as the antenna filter [6–11], balun filter [12–15], phase-shifter filter, antenna package [16,17], and antenna circulator [18,19], have been reported. In this chapter we concentrate on a discussion of the first three.

Compact Multifunctional Antennas for Wireless Systems, First Edition. Eng Hock Lim, Kwok Wa Leung.
© 2012 John Wiley & Sons, Inc. Published 2012 by John Wiley & Sons, Inc.

2.2 ANTENNA FILTERS

Antennas and filters are among the components that are very difficult to miniaturize, as they may require the use of resonators, which are usually bulky. The antenna filter is a newly proposed component, first reported by Lim and Leung [6], that combines the radiating and filtering elements into a single module. In general, a simple antenna filter requires at least three ports, with the simplest case shown in Fig. 2.1(a). It can easily be extended into a multiport configuration, with the signal flows as illustrated in Fig. 2.1(b). This multifunctional component can be designed either with a single resonator or with a combination of multiple resonators. Designing a good antenna filter requires meeting the following criteria:

1. Small in size and easy to design
2. Good performance from both the individual antenna (such as wide impedance bandwidth, good polarization, and high directivity and radiation efficiency)

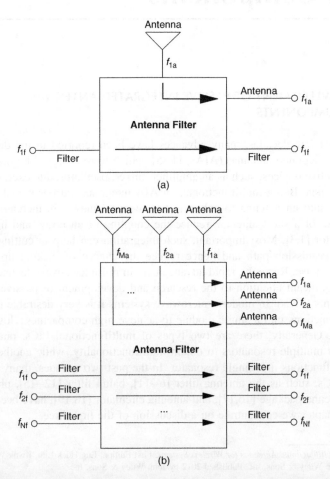

FIGURE 2.1 Configurations of two antenna filters: (a) three-port; (b) multiport.

and filter (such as high selectivity, broad rejection frequency bandwidth, and low spurious response)

3. Good isolation between the antenna and filter parts
4. High degree of freedom to design the operating frequencies of the antenna and filter arbitrarily and separately

Modeling an antenna filter usually requires the use of field-based simulation tools so that the coupling mechanism between the antenna and filter parts can be accounted for. In this chapter the design ideas are demonstrated using dielectric and microstrip resonators.

2.2.1 Dielectric Resonator Antenna Filter

Traditionally, a dielectric resonator (DR) has been used for filter and oscillator designs because of its high Q factor [20,21]. The most distinctive advantage of a DR is that it does not have any metallic loss. A DR can be made into a variety of shapes, such as cylindrical, hemispherical, or rectangular. Some common DR examples are shown in Figs. 2.2 and 2.3. A high Q factor is very essential for achieving high frequency selectivity in a filter and low phase noise in an oscillator. The simplest excitation scheme for a DR filter (DRF) and a DR oscillator (DRO) is to use the proximity side-coupled method shown in Fig. 2.4. In designing a DRF or a DRO, an external metallic enclosure is generally used to reduce the radiation

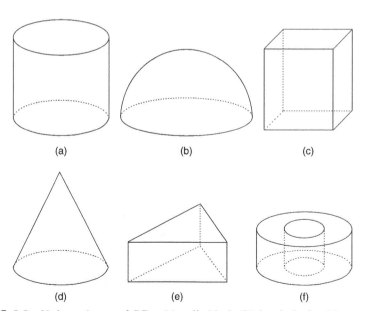

FIGURE 2.2 Various shapes of DRs: (a) cylindrical; (b) hemispheric; (c) rectangular; (d) conical; (e) triangular; (f) annular ring.

FIGURE 2.3 Common DR samples.

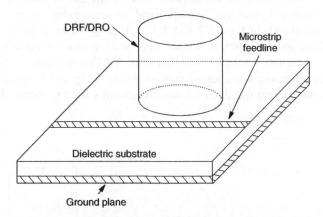

FIGURE 2.4 Proximity side-coupled excitation scheme for a DRF or a DRO.

loss, at the cost of increasing the component size. A commercial DRO is shown in Fig. 2.5. Obviously, the enclosure is much larger than the DR.

It was first shown by Long et al. in 1983 [22] that a DR could also be used as an effective antenna. Since then, the DR antenna (DRA) has received considerable attention because of a number of advantages, such as its small size, low loss, low cost, and light weight. The excitation of a DRA is very straightforward. It is worth mentioning that most of excitation methods for microstrip antennas are usually applicable to DRAs. Several popular DRA excitation schemes are displayed in Fig. 2.6. As for the coaxial-probe feeding method shown in Fig. 2.6(a), the probe can be placed either internally (position A) or externally (position B) to the DR. The metallic probe may generate extra ohmic loss and self-reactance at millimeter-wave frequencies, which are very undesirable. Also, the internal probing requires

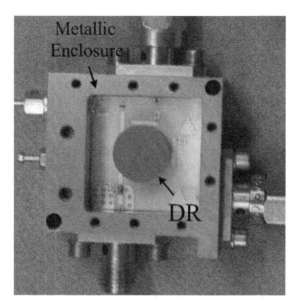

FIGURE 2.5   Commercial DRO.

drilling a hole in the superhard DR, which is very tough in reality. As a result, the microstrip- and aperture-fed schemes [Fig. 2.6(b) and (c)] are more attractive since a DR can easily be excited simply by placing it on top of an excitation source. It is worth mentioning that the two microstrip-oriented feeding methods can easily be integrated with monolithic microwave integrated circuits. In practice, however, for all the aforementioned excitation schemes, the formation of an airgap between the DR and the ground plane is sometimes unavoidable during the manufacturing process, and it can seriously affect the coupling efficiency of the fields. To overcome this problem, the conformal-strip feeding method was proposed, in which a metallic strip is closely attached on the DR surface, as depicted in Fig. 2.6(d). Noticeably, this method shares most of the principal advantages of the coaxial-probe method.

In the past few years, many reports have been published on DRAs [22–24] and DRFs [25–28]. Thus far, however, very little or no work has been carried out to combine a DRA and a DRF using a single piece of DR. The Q factors of the DRF and DRO are usually high, to reduce the insertion loss, but that of the DRA is usually low, to enhance the radiation and bandwidth. Because of the very different Q-factor requirements, it seems intuitively contradictory to make a DRA and a DRF using a single piece of DR. The first successful demonstration of a DRA filter (DRAF), a dual-functional device that combines an antenna and a filter in a single piece, was reported in 2008 [6]. In this newly proposed dual-functional component, the antenna and filter share the same resonator. In this case, two different resonances of a resonator are used in designing the two very different functions. Use is made of the $TE_{01\delta}$ and $HEM_{11\delta}$ modes of the DR to design the DRF and DRA, respectively. Because of the orthogonality of the two modes, the antenna and filter parts of the DRAF can be designed and tuned separately and almost independently. It was found that the antenna and filter parts can also be designed at the same or different frequencies. A cylindrical DR loaded with a metallic disk is utilized as the resonator for the antenna as well as for the bandpass filter. A top-loading metallic disk is used to improve the insertion loss of the DRF and simultaneously, to tune the filter. It has a negligible effect on the radiation efficiency of the DRA.

Configuration   The design idea is demonstrated using a cylindrical DR with a radius R_d of 10.5 mm, a height H of 9 mm, and a dielectric constant ε_r of 34. The $TE_{01\delta}$ and $HEM_{11\delta}$ modes of a bare cylindrical DR can be calculated using the empirical equations (2.1) and (2.2) [23].
$TE_{01\delta}$ mode:

$$f_0 = \frac{2.921c\varepsilon_r^{-0.465}}{2\pi R_d}\left[0.691 + 0.319\left(\frac{R_d}{2H}\right) - 0.035\left(\frac{R_d}{2H}\right)^2\right] \quad (2.1)$$

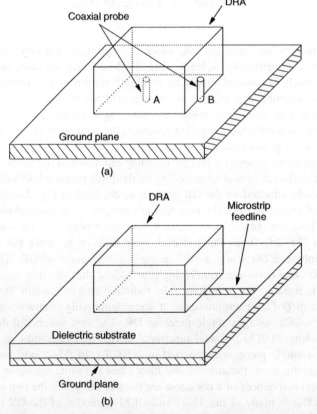

FIGURE 2.6   DRA with various feeding schemes: (a) coaxial probe; (b) microstripline; (c) microstrip-fed aperture; (d) vertical strip.

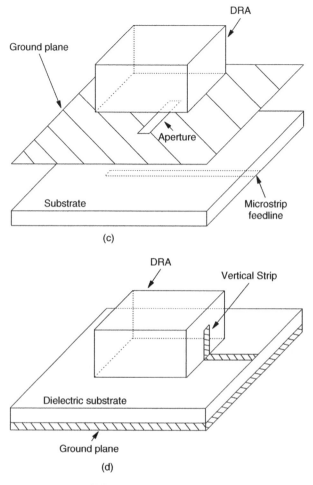

FIGURE 2.6   (*Continued*)

HEM$_{118}$ mode:

$$f_0 = \frac{2.735 c \varepsilon_r{}^{-0.436}}{2\pi R_d}\left[0.543 + 0.589\left(\frac{R_d}{2H}\right) - 0.050\left(\frac{R_d}{2H}\right)^2\right] \quad (2.2)$$

where c is the speed of light in air. The TE$_{018}$ and HEM$_{118}$ field patterns of a cylindrical DR are sketched in Fig. 2.7. When used for a DRF design with a ground plane, the DR is half-truncated (cut along the line GG′ to remove the bottom-half DR). Usually, two 50-Ω microstrips are placed adjacent to the DR to excite the magnetic field of the end-fire TE$_{018}$ mode. Similarly, a centrally placed microstrip (at the maximum electric field point) can be used to excite the broad-side HEM$_{118}$ mode in the half-volume DR for the antenna. For ease of reference, the fields of

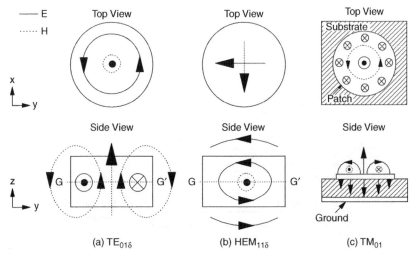

FIGURE 2.7   Electric and magnetic field patterns for (a) the $TE_{01\delta}$ mode of a cylindrical DR; (b) the $HEM_{11\delta}$ mode of a cylindrical DR; (c) the TM_{01} mode of a cylindrical microstrip patch antenna. (From [6], copyright © 2008 IEEE, with permission.)

the dominant mode of a circular microstrip patch are also given in Fig. 2.7. With reference to Fig. 2.7(c), it is obvious that there are only TM modes in the patch, due to the boundary condition of the magnetic field, $H_z = 0$ [29,30]. Owing to this limitation, the microstrip patch cannot be used to realize this idea.

Figure 2.8 shows the configuration of the first-order DRAF, where the afore-mentioned DR is used. A circular metallic disk of radius $R_c = 10$ mm is displaced concentrically at $h = 4$ mm from the top of the DR. HFSS simulations show that the DRAF is working in its fundamental broad-side $HEM_{11\delta}$ mode at 2.68 GHz [23]. A DRAF prototype is shown in Fig. 2.9.

It is well known that the fundamental end-fire $TE_{01\delta}$ mode can be excited in a cylindrical DR when the condition $R_d > H$ [20] is met. HFSS simulation shows that the resonance frequency of the $TE_{01\delta}$ mode of the first-order DRAF is 3.095 GHz. In this case, two conventional 50-Ω microstrips ($w = 1.99$ mm) are etched orthog-onally on a Duroid substrate ($\varepsilon_{rs} = 2.94$ and thickness 0.762 mm) to side-couple the DR. As can be seen from Fig. 2.8, microstrips 1 and 2 are used to excite the $TE_{01\delta}$ mode of the DR for designing the filter part of the DRAF. The feedlines have an offset of $D = 11.5$ mm from the center of the DR, with matching stub lengths of $L_{s1} = L_{s2} = 10$ mm. For the antenna operation, the DR is placed on top of a feedline for excitation of the broad-side $HEM_{11\delta}$ mode. The microstrip has a matching stub offset of $L_{s3} = 4.6$ mm, with an inclination angle of $\phi' = 45°$.

A mode chart shows the effect of a design parameter on a certain resonance. Figure 2.10 shows a simulated mode chart for the fundamental end-fire $TE_{01\delta}$ and broad-side $HEM_{11\delta}$ modes of a microstrip-fed cylindrical DR as a function of R_d/H, with $R_d = 10.5$ mm and $\varepsilon_r = 34$. The HFSS simulation models (50-Ω microstrip feedlines, with substrate $\varepsilon_{rs} = 2.94$ and thickness 0.762 mm) are also

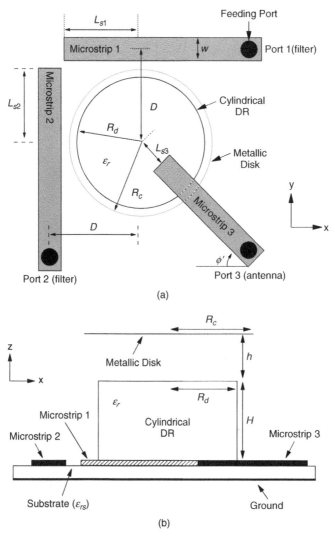

FIGURE 2.8 First-order cylindrical DRAF: (a) top view; (b) front view. (From [6], copyright © 2008 IEEE, with permission.)

shown in the insets, where the side- and centrally coupling methods are used to excite the $TE_{01\delta}$ and $HEM_{11\delta}$ modes, respectively. As can be seen from the figure, the resonance frequencies of both modes increase with increasing R_d/H ratio in the frequency range 2.5 to 5.5 GHz.

Results and Discussion The commercial software Ansoft HFSS was used to simulate a DRAF, and measurements were carried out to verify the simulations. Usually, simulations that involve three-dimensional microwave configurations and

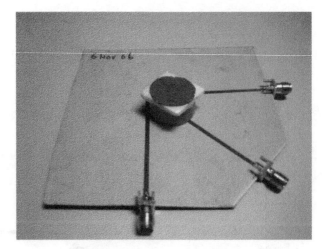

FIGURE 2.9   First-order DRAF.

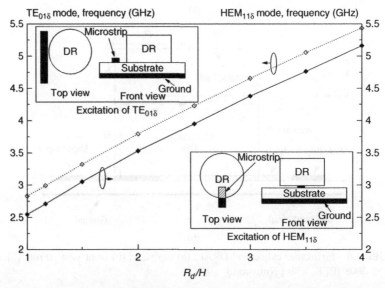

FIGURE 2.10   Mode chart for the fundamental end-fire $TE_{01\delta}$ and broad-side $HEM_{11\delta}$ modes of a microstrip-fed cylindrical DR, with $R = 10.5$ mm. The insets show the top and side views of the HFSS simulation models. (From [6], copyright © 2008 IEEE, with permission.)

radiation computations are done more conveniently using field-based software tools. In microwave measurements, all the unused ports were terminated by 50-Ω loads.

The first-order DRAF in Fig. 2.8 was studied first. The simulated and measured reflection coefficients, insertion losses, and mutual couplings between ports are shown in Fig. 2.11. The results of the filter part (DRF) of the DRAF are discussed

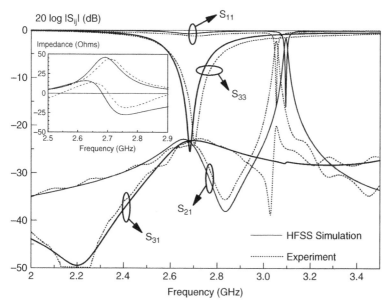

FIGURE 2.11 Simulated and measured reflection coefficients, insertion losses, and coupling coefficients of the first-order cylindrical DRAF (Fig. 2.8). The inset shows the simulated and measured input impedances of the DRA. (From [6], copyright © 2008 IEEE, with permission.)

first. As a side-coupling scheme is used for exciting the DR, it was found from simulations that a high dielectric constant is needed for efficient field coupling. A higher dielectric constant, say $\varepsilon_r > 20$, is needed to excite the $TE_{01\delta}$ mode efficiently. Also, it is found from simulations that the $TE_{01\delta}$ mode cannot be excited properly when a DR of $\varepsilon_r = 10$ is used. As can be seen from Fig. 2.11, the measured and simulated operating frequencies of the DRF are 3.054 and 3.095 GHz, respectively, with an error of 1.34%. The insertion loss measured is 2.19 dB at the operational frequency, which is slightly larger than that for simulation (1.6 dB). This can be caused by omission of the SMA connectors in the simulation. Also, for a simplified simulation model, perfect conductors and lossless dielectrics were used to simulate the feedlines and substrates. In practice, each SMA connector has an insertion loss of about 0.15 to 0.2 dB over the frequency range, and the conductor has some additional loss, which can be quite severe in millimeter-wave ranges. As can be seen in Fig. 2.11, the measured and simulated 3-dB passbands of the DRF are about 18 and 17 MHz, respectively. With reference to Fig. 2.11, both the measured and simulated reflection coefficients at port 1 (from S_{11}) are about 16 dB at the resonance. Similar results were obtained for the reflection coefficients at port 2, which is expected because of the symmetry.

Next, the antenna (DRA) part (port 3 in Fig. 2.8) of the DRAF is analyzed. The measured and simulated resonance frequencies are given by 2.70 and 2.68 GHz (with an error of 0.74%), respectively. With reference to Fig. 2.11, the measured and simulated impedance bandwidths ($S_{33} \leq -10$ dB) are 3.68 and 3.35%, respectively. The measured and simulated input impedances of the DRA are shown in the inset, with good agreement observed between the simulation and the measurement. The coupling between the DRA and DRF ports (S_{31}, S_{32}) is in general less than -20 dB across the entire frequency range. This shows that the antenna and filter parts can be designed and tuned almost independently.

The measured and simulated radiation patterns of the antenna part (port 3 in Fig. 2.8) of the DRAF are illustrated in Fig. 2.12. The E- and H-planes are defined at the cut planes of $\phi' = 45°$ and $-45°$, respectively, as shown in Fig. 2.8. As can be seen from the figure, a broad-side $\text{HEM}_{11\delta}$ mode is obtained, which is expected for the fundamental mode. In the broad-side direction ($\theta = 0°$), the cross-polarized fields are weaker than their co-polarized counterparts by at least 20 dB. Figure 2.13 shows the measured antenna gain of the DRA as a function of frequency. It can be seen from the figure that the antenna gain is approximately 5 dBi at resonance. The measured antenna gain with the metallic disk removed is shown for comparison. It is found that the metallic disk has little effect on the antenna gain. The simulated radiation patterns for the two configurations (DRAFs with and without the metallic disk) are compared in Fig. 2.14. As can be seen from the figure, the patterns of the two configurations are quite similar. It shows that the top-loading metallic disk can be used to tune the filter and improve filter performance without greatly affecting the normal operation of the antenna.

For an antenna filter, it is very important that the antenna and the filter can be designed separately. This is possible if the coupling between the two elements is

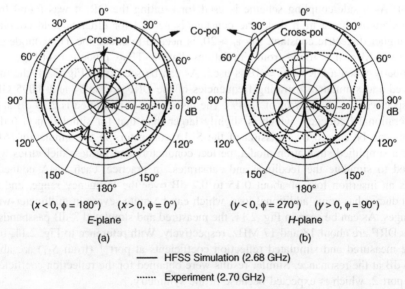

FIGURE 2.12 Simulated and measured normalized radiation patterns of the first-order cylindrical DRAF (Fig. 2.8): (a) E-plane; (b) H-plane. (From [6], copyright © 2008 IEEE, with permission.)

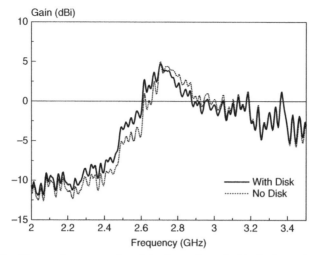

FIGURE 2.13 Measured antenna gains of the first-order cylindrical DRAFs in Fig. 2.8, with and without a loading metallic disk. (From [6], copyright © 2008 IEEE, with permission.)

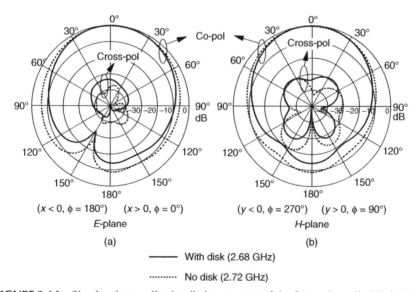

FIGURE 2.14 Simulated normalized radiation patterns of the first-order cylindrical DRAF in Fig. 2.8: (a) E-plane; (b) H-plane. (From [6], copyright © 2008 IEEE, with permission.)

small enough. To investigate further, an isolated DRA (without the filter part) was constructed by removing microstrips 1 and 2 (in Fig. 2.8). Similarly, an isolated DRF was built by removing microstrip 3 from the DRAF. Table 2.1 compares the simulated resonance frequencies of the stand-alone DRA or DRF with those of the DRAF. As can be seen from the table, the resonance frequencies of the antenna

TABLE 2.1   Comparison of Simulated Resonance Frequencies (GHz) of a DRAF and an Isolated DRA/DRF

	Antenna Frequency	Filter Frequency
DRAF	2.68	3.095
Isolated DRA or DRF	2.66	3.106
	(isolated DRA)	(isolated DRF)

and filter parts of the DRAF are very close to those of the stand-alone DRA or DRF. This is very encouraging, as it shows that designs of the antenna and filter parts can be done quite independently for the DRAF, due to the orthogonality of the electric fields of the $TE_{01\delta}$ and $HEM_{11\delta}$ modes, which can be seen clearly from Fig. 2.7(a) and 2.7(b).

Next, we discuss the performances of the filter part of a DRAF. Table 2.2 compares the loaded Q factor (Q_L) and out-of-band rejection of the DRAF with those of existing bandpass filters. The Q_L factor is calculated using $Q_L = f_0/BW_{3dB}$, where f_0 and BW_{3dB} are the measured central operating frequency and 3-dB filter bandwidth [31], respectively. As can be seen from the table, the DRAF has a much higher Q factor than that for conventional microstrip-based bandpass filters, which have a higher conductor loss. Although the Q factor of the DRAF is much lower than for the conventional cavity-encapsulated DRF [32], the DRAF has the advantage of being more compact. The Q_L factor of a conventional cavity-encapsulated DRF can easily be made greater than 1000. A metallic cavity cannot be used in this case as it disturbs the antenna operation. Of course, the DRAF can be further miniaturized by increasing the permittivity, but this is not the focus here. The out-of-band-performance is found to be greater than 30 dB.

Parametric studies have been performed on the top-loading metallic disk. Figure 2.15(a) shows the effect of the height h on the operating frequencies of a DRA and a DRF. As can be seen from the figure, the DRA frequency increases from 2.65 GHz to 2.72 GHz (2.64%), and the DRF frequency decreases from 3.23 GHz to 3.00 GHz (7.12%), as h is increased from 2 to 14 mm. The effects of

TABLE 2.2   Comparison of the DRAF and Existing Bandpass Filters

Type of Narrowband Bandpass Filter	Frequency (GHz)	Largest Dimension (mm)	Q Factor, Q_L	Out-of-Band Rejection (dB)
Filter part of the DRAF	3.054	21	169.66	~34
Microstrip square ring	2	27	104.45	
Microstrip patch	1.6	20	50	~25
Co-planar waveguide	2.965	21.4	14.363	~50
Electromagnetic bandgap	9	29.1	40	~25

Source: [6].

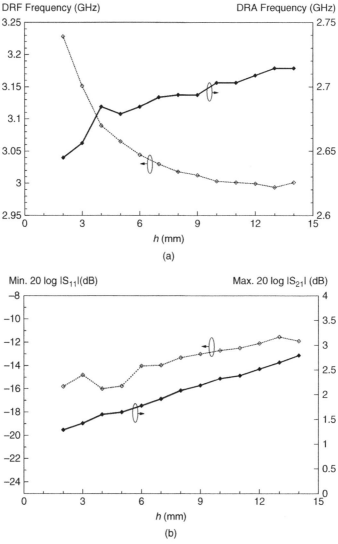

FIGURE 2.15 Effect of the height h of a metallic disk on the DRAF in Fig. 2.8: (a) resonance frequencies of the DRA and the DRF as a function of h; (b) minimum 20 log $|S_{11}|$ (maximum return loss) and maximum 20 log $|S_{21}|$ (minimum insertion loss) as a function of h. (From [6], copyright © 2008 IEEE, with permission.)

h on the reflection coefficients and insertion losses are shown in Fig. 2.15(b). It is obvious that both the reflection coefficients and insertion losses can be improved by using a smaller h. Figure 2.16 shows the effect of the disk radius R_c on the DRAF. As can be observed from Fig. 2.16(a), the DRA frequency decreases with, and the DRF frequency increases with, increasing R_c. As can be seen from Fig. 2.16(b), both the reflection coefficients and insertion losses can be improved

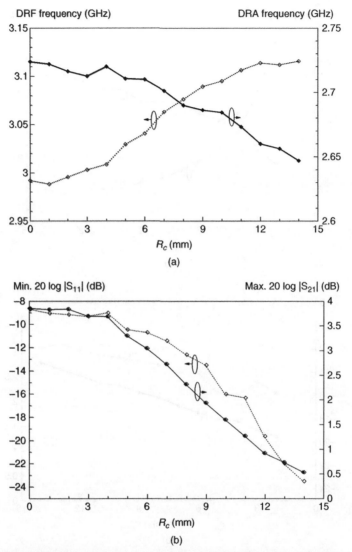

FIGURE 2.16 Effect of the radius R_c of a metallic disk on the DRAF in Fig. 2.8: (a) resonance frequencies of the DRA and the DRF as a function of R_c; (b) minimum 20 log $|S_{11}|$ (maximum return loss) and maximum 20 log $|S_{21}|$ (minimum insertion loss) as a function of R_c. (From [6], copyright © 2008 IEEE, with permission.)

by using a larger disk. Obviously, the DRAF can be fine-tuned by varying the height and radius of the metallic disk, which can be a very important feature for the filter [33,34].

The DRAF in Fig. 2.8 has different antenna and filter operating frequencies. It will be shown that a DRAF can also be designed to have the two frequencies coincident with each other. Figure 2.17 shows the simulated results, with parameters

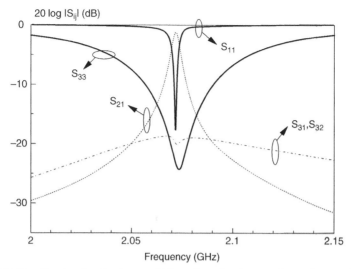

FIGURE 2.17 Simulated reflection coefficient, insertion loss, and coupling coefficient of a disk-loaded cylindrical DRAF shown as a function of frequency. In this case, a DRA and a DRF have the same operating frequency. (From [6], copyright © 2008 IEEE, with permission.)

$\varepsilon_r = 80$, $R_d = 10$ mm, $H = 8.8$ mm, $R_c = 10$ mm, $h = 5.2$ mm, $D = 11.8$ mm, $L_{s1} = L_{s2} = 10$ mm, and $L_{s3} = 2.2$ mm. It should be mentioned that although both the DRA and DRF operate at the same frequency, the coupling between them (S_{31}, S_{32}) is still very small, about -20 dB. This implies that the antenna and filter parts can be designed separately.

A second-order DRAF is also designed using two disk-loaded DRs, with the configuration shown in Fig. 2.18. The optimized design parameters are $\varepsilon_r = 34$, $R_d = 10.5$ mm, $H = 9$ mm, $R_c = 10$ mm, $h = 4$ mm, $D = 12$ mm, $d = 34$ mm, $L_{s1} = L_{s2} = 16$ mm, and $L_{s3} = L_{s4} = 4.6$ mm. All of the microstrip feedlines are 50-Ω lines. Ports 1 and 2 are for the DRF, whereas ports 3 and 4 are for the DRA. Figure 2.19 shows the simulated reflection coefficients, insertion losses, and coupling coefficients of a second-order DRAF. During the tuning process, it was found that the resonance frequencies of the antenna and filter parts were almost independent of each other. This is not surprising, due to the orthogonality between the fields of the modes. Furthermore, microstrips 1 and 2 (filter ports) are normal to microstrips 3 and 4 (antenna ports), making the coupling between the microstrip lines very small. With reference to Fig. 2.19, the simulated DRA frequency at port 3 (or port 4) is 2.71 GHz, which is only slightly higher than for the first-order version. Two modes are observed for the DRF, which is expected for a dual-mode bandpass filter. The out-of-band rejection of the filter part of the second-order DRAF is much better than that of its first-order counterpart. Figure 2.20 shows the simulated E-plane (yz-plane) and H-plane (xz-plane) radiation patterns of the antenna part of the second-order DRAF. Again, the co-polarized fields are stronger than their

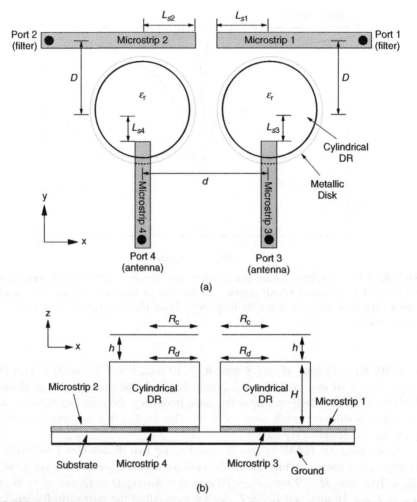

FIGURE 2.18   Second-order DRAF: (a) top view; (b) side view. (From [6], copyright © 2008 IEEE, with permission.)

cross-polarized counterparts by at least 20 dB in the broad-side direction. It should be mentioned that a higher-order DRAF can easily be obtained by increasing the number of DRs.

2.2.2   Other DRAFs

Other DRAFs have also been reported [7,8]. A dual-functional DR that works simultaneously as a diversity DRA and a bandpass DRF was reported by Hady et al. [7]. The configuration is shown in Fig. 2.21. It can be seen that two resonances ($TM_{01\delta}$ and $HEM_{11\delta}$) of the DR are excited at ports 3 and 4 for the antenna. At the same time, the high-Q $TE_{01\delta}$ has been excited for a bandpass DRF at ports 1

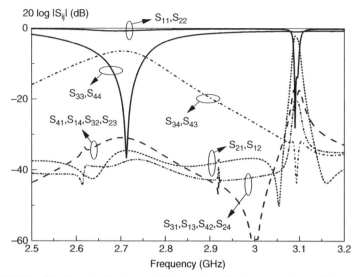

FIGURE 2.19 Simulated reflection coefficients, insertion losses, and coupling coefficients of the second-order DRAF (Fig. 2.18) shown as a function of frequency. (From [6], copyright © 2008 IEEE, with permission.)

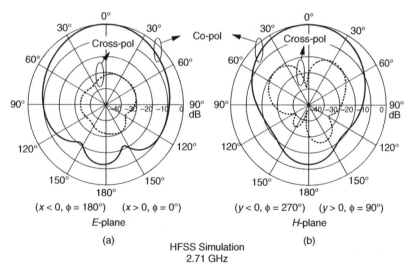

FIGURE 2.20 Simulated normalized radiation patterns of the second-order DRAF in Fig. 2.18: (a) E-plane; (b) H-plane. (From [6], copyright © 2008 IEEE, with permission.)

and 2. A Trans-Tech DR with $\varepsilon_{r,DR} = 36.6$, $R_{DR} = 13.27$ mm, and $H = 8.33$ mm was used for the experiment. It is excited by a circular metallic patch being fed by the center conductor of an SMA connector, shown in Fig. 2.21. The microstrip feeding networks are built on two pieces of Duroid 6002 ($\varepsilon_{rs} = 2.94$, $W_{sub} = 80$ mm, and $t = 1.524$ mm) stacked bottom to bottom, with a ground plane inserted

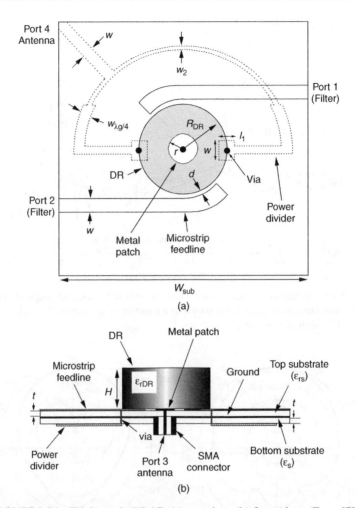

FIGURE 2.21 Triple-mode DRAF: (a) top view; (b) front view. (From [7].)

at the center. For the filter part, two curved microstrip feedlines are placed adjacent to the DR for excitation of the $TE_{01\delta}$ mode. A power divider, placed at the bottom surface of the bottom substrate, is used to obtain the broad-side $HEM_{11\delta}$ mode in the DRA. The $TM_{01\delta}$ mode, being end-fire in nature, is excited by a circular metallic patch placed at the bottom of the DR. Other design parameters are $w = 3.788$ mm, $w_2 = 1.014$ mm, $w_{\lambda_g/4} = 1.45$ mm, $r = 6$ mm, $d = 1.1$ mm, and $l_1 = 3$ mm. Further information and results may be found in reference [7].

The same excitation scheme (that was used by Hady et al. [7]) was also used for exciting the $TM_{01\delta}$ mode in a cylindrical DR for antenna design [8]. A coaxial-fed circular patch (with $r = 3$ mm) was used here. In the same piece of work, a conventional side-coupled microstrip was used to excite the high-Q $TE_{01\delta}$ mode to implement a DRF. In this case the DR is placed 1.5 mm away from the edge of the microstrip to ensure efficient coupling. The configuration is shown in Fig. 2.22. The same DR and substrate were used again for this new design. A cavity (49 × 53 × 25 mm^3) made of metallic strips was used to improve the insertion loss of the filter. The strips were aligned horizontally so that they were virtually invisible to the antenna mode. The horizontal strips were made on the sidewalls of a thin RT/Duroid 6002 substrate with a strip width of 3 mm and the spacing width of 2 mm (between strips). In addition, a top cover that is made of circular strips (with a strip width of 2 mm and spacing width of 3 mm) is placed on top of the cavity. Other detailed results are available in reference [8].

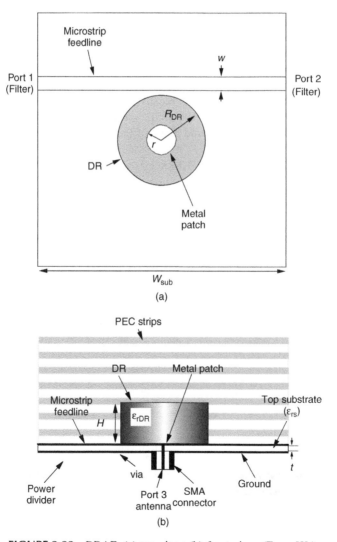

FIGURE 2.22 DRAF: (a) top view; (b) front view. (From [8].)

2.2.3 Microstrip-Based Antenna Filters

Rectangular Microstrip Ring–Patch Antenna Filter In this section a three-port antenna filter that is made on a single substrate is studied [9]. By integrating the antenna and filter together, the system size can be reduced considerably. This multifunction module consists of a microstrip ring and a microstrip patch, which can be placed very close to each other. It was found that the coupling between the antenna and filter is low, so it is not necessary to use isolating ground layers. The proposed design is particularly useful for system-on-package solutions, where the antenna and filter parts are usually made separately on the top surface of a package [35–37].

Configuration The configuration of the microstrip antenna filter proposed [9] is illustrated in Fig. 2.23, where a probe-fed rectangular patch antenna ($d_2 = 35$ mm and $g_2 = 4.3$ mm) is placed concentrically in the central area of a rectangular ring filter ($d_1 = 54$ mm, $w = 1.99$ mm, and $g_1 = 0.2$ mm). In this case, the filter part has a narrow bandwidth. A Duroid substrate of $\varepsilon_r = 2.94$ and thickness $h = 0.762$ mm was used in the design here. Typically, the unloaded Q factor of the ring-shaped microstrip filter is about 100 [38]. Since the Q factor is high, the filter response is narrowband. The ground plane of the antenna filter is 12 cm × 12 cm.

Results and Discussion Ansoft HFSS was used to study the reflection coefficient, radiation pattern, insertion loss, and mutual coupling of the antenna filter. In the measurements, each unused port was terminated with a 50-Ω load. Each

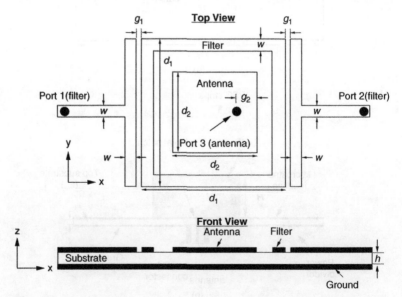

FIGURE 2.23 Antenna filter consisting of a rectangular microstrip patch antenna and a ring filter. (From [9], copyright © 2010 IEEE, with permission.)

configuration was fabricated on a single substrate. The simulations were verified by measurements, and good agreement between them was obtained. Figure 2.24 shows the simulated and measured reflection coefficients, insertion losses, and mutual couplings between different ports. The filter performances (ports 1 and 2) of the antenna filter are first studied. As can be seen from the figure, the measured and simulated resonance frequencies of the filter part are 930.500 and 917.375 MHz, respectively, with an error of 1.43%. The measured and simulated insertion losses are 2 and 0.082 dB, respectively. A higher loss is found in the measurement because it has included the losses of the feedlines and SMA connectors used in the experiment. With reference to the figure, the measured and simulated 3-dB passbands are 25 and 29 MHz, respectively. At port 1, the measured and simulated reflection coefficients are 15.30 and 21.27 dB, respectively. Similar results were obtained for port 2, which is expected because of the symmetry. A higher-order mode is found at 1.90 GHz, which was also reported by Chang and Hsieh [38].

The performances of the antenna part (port 3) of the antenna filter are now addressed. The patch antenna operates in its fundamental TM_{01} mode. With reference to Fig. 2.24, the measured and simulated resonance frequencies are 2.437 and 2.395 GHz, respectively, with an error of 1.75%. The measured and simulated impedance bandwidths ($S_{33} \leq -10$ dB) are 0.81 and 0.54%, respectively. This is typical, as the Q factor for the patch resonator is usually high. It is noted from the figure that the coupling between the antenna and filter ports is generally weaker than 40 dB over the entire frequency range. This is highly desirable, as it enables the filter and antenna parts to be designed independently. The measured and simulated

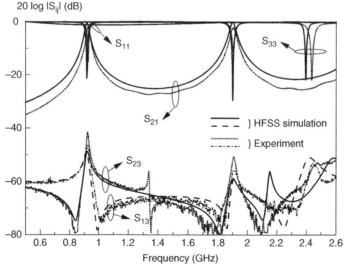

FIGURE 2.24 Simulated and measured reflection coefficients, insertion losses, and coupling coefficients of the antenna filter shown as a function of frequency in Fig. 2.23. (From [9], copyright © 2010 IEEE, with permission.)

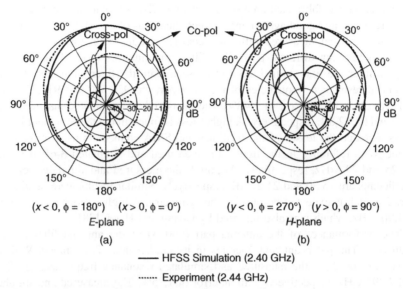

$(x < 0, \phi = 180°)$ $(x > 0, \phi = 0°)$

E-plane

(a)

$(y < 0, \phi = 270°)$ $(y > 0, \phi = 90°)$

H-plane

(b)

—— HFSS Simulation (2.40 GHz)

······ Experiment (2.44 GHz)

FIGURE 2.25   Simulated and measured normalized radiation patterns of the antenna filter in Fig. 2.23. (From [9], copyright © 2010 IEEE, with permission.)

radiation patterns are given in Fig. 2.25. As can be observed from the figure, a broad-side mode is obtained [39], with the cross-polarized fields weaker than the copolarized fields by at least 18 dB in the bore-sight direction ($\theta = 0°$). The measured cross-polarized fields are generally stronger than the simulated fields, due to various experimental tolerances. Referring to Fig. 2.26, the measured antenna gain is 4.24 dBi at 2.44 GHz, which is lower than that for a typical microstrip antenna.

Circular Microstrip Ring–Patch Antenna Filter

Configuration   Figure 2.27 shows a microstrip antenna filter that is designed by combining a circular patch and an open-ring resonator [10] into a single resonator. Unlike in the previous section, the antenna and filter here are made on a single resonator. As can be seen from the figure, the patch is placed inside the ring for a smaller footprint. Ports 1 and 2 are the filter ports, and port 3 is for feeding the microstrip patch. The antenna and filter are interconnected by a high-impedance line to minimize their interaction. It will be shown later that the isolation between both components can be made very high. The optimized design parameters are given here: $R_1 = 22.2$ mm, $R_2 = 21.8$ mm, $R_3 = 15$ mm, $w_1 = 0.5$ mm, $w_2 = 0.2$ mm, $s = 3$ mm, $d = 0.5$ mm, $g = 0.3$ mm, and $\beta = 30°$. Other parametric analyses are available [10].

Results and Discussion [10]   Simulations were carried out by Sung using IE3D software, and experiments were also conducted. Figure 2.28 shows the simulated and measured S parameters of the microstrip ring-patch antenna filter. The filter part is discussed first. With reference to the figure, the center operating frequency

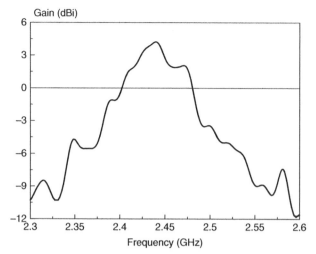

FIGURE 2.26 Measured antenna gain of the antenna filter in Fig. 2.23. (From [9], copyright © 2010 IEEE, with permission.)

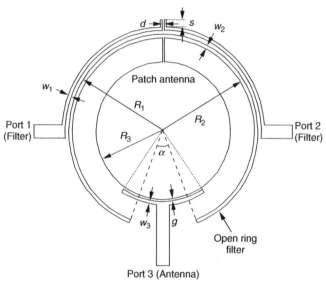

FIGURE 2.27 Antenna filter consisting of a circular microstrip patch antenna and a ring filter. (From [10], copyright © 2009 IEEE, with permission.)

of the filter is about 0.67 GHz and the 3-dB bandwidth is about 15.1%. There is a minimum insertion of 1.5 dB in the filter passband. Two transmission zeros can be observed at 0.52 and 1.14 GHz, which have significantly improved the frequency selectivity. It is worth mentioning that the coupling (S_{23}) between the antenna and filter ports is well below 25 dB in the frequency range 1 to 3 GHz.

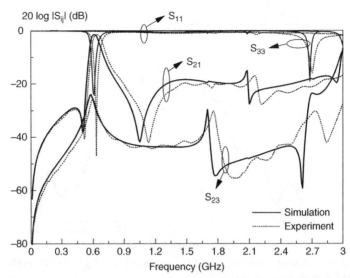

FIGURE 2.28   Simulated and measured reflection coefficients, insertion losses, and coupling coefficients of the antenna filter shown as a function of frequency. (Courtesy of Y. J. Sung, Kyonggi University. From [10], copyright © 2009 IEEE, with permission.)

It shows that the two resonators have very little interaction and can be designed almost independently.

Next, the antenna part is studied. As can be seen from Fig. 2.28, the antenna has a resonance frequency of 2.7 GHz. The antenna bandwidth measured is about 1.7%, being higher than that for simulation (0.8%). The discrepancy can be caused by the omission of conductor and dielectric losses in the simulation. Figure 2.29 shows the electric current distribution on the conductor surface of the filter at 2.7 GHz, showing good isolation between the ports. The measured radiation patterns are shown in Fig. 2.30. As can be seen from the figure, the co-polarized fields are larger than their cross-polarized counterparts in both the E- and H-planes.

Microstrip Ring–Slot Antenna Filter

Configuration   Figure 2.31 displays the configuration of the microstrip ring–slot antenna filter, where a wideband slot antenna ($d_3 = 35$ mm, $d_4 = 10$ mm, $d_5 = 5$ mm, $w_1 = 1.99$ mm, $w_2 = 10$ mm, $w_3 = 17$ mm, $w_4 = 16.5$ mm, $g_3 = 0.51$ mm, and $g_4 = 1$ mm) is etched in the central part of a narrowband U-shaped microstrip filter ($d_1 = 40$ mm, $d_2 = 35$ mm, $g_1 = 0.2$ mm, and $g_2 = 21.005$ mm).

Results and Discussion   The simulated and measured reflection coefficients, insertion losses, and mutual couplings between different ports are shown in Fig. 2.32. First, the filter part is discussed. As can be seen from the figure, the measured and simulated resonance frequencies are 903.125 and 900.000 MHz (with an error of 0.35%), respectively. The measured and simulated insertion losses are 2.53 and 0.28 dB. It is observed that the measured 3-dB passband is about 20 MHz, which is

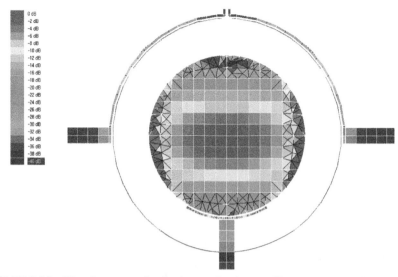

FIGURE 2.29 Electric current distribution on the antenna filter at resonance in Fig. 2.27. (Courtesy of Y. J. Sung, Kyonggi University. From [10], copyright © 2009 IEEE, with permission.)

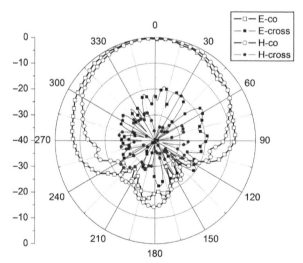

FIGURE 2.30 Measured normalized radiation patterns of the antenna part in Fig. 2.27. (Courtesy of Y. J. Sung, Kyonggi University. From [10], copyright © 2009 IEEE, with permission.)

slightly narrower than the simulated value of 22 MHz. The measured and simulated reflection coefficients (S_{11}) at port 1 are 15.8 and 19.1 dB, respectively. It is worth mentioning that the higher-order mode of the filter (about 1.8 GHz) is suppressed well below 30 dB.

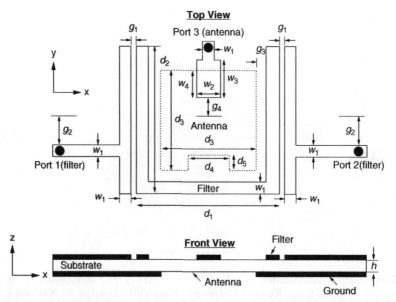

FIGURE 2.31 Antenna filter consisting of a slot antenna and a ring filter. (From [9], copyright © 2010 IEEE, with permission.)

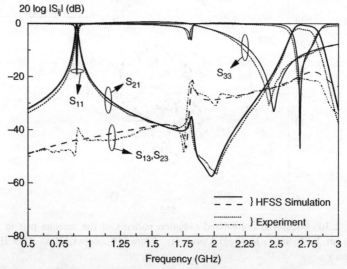

FIGURE 2.32 Simulated and measured reflection coefficients, insertion losses, and coupling coefficients of the antenna filter shown as a function of frequency. (From [9], © copyright 2010 IEEE, with permission.)

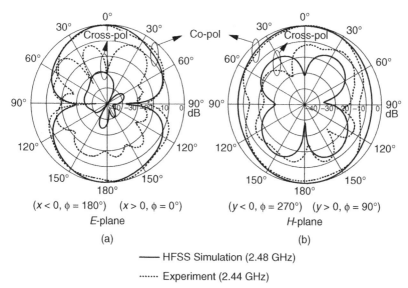

(x < 0, φ = 180°)   (x > 0, φ = 0°)   (y < 0, φ = 270°)   (y > 0, φ = 90°)

E-plane   H-plane

(a)   (b)

——— HFSS Simulation (2.48 GHz)

······· Experiment (2.44 GHz)

FIGURE 2.33   Simulated and measured normalized radiation patterns of the antenna part of the antenna filter. (From [9], copyright © 2010 IEEE, with permission.)

The antenna part of an antenna filter is investigated next. With reference to Fig. 2.32, the measured and simulated resonance frequencies at port 3 (S_{33}) are given as 2.441 and 2.481 GHz (with an error of 1.7%), respectively. The measured and simulated impedance bandwidths ($S_{33} \leq -10$ dB) are 24 and 20.65%. A small resonance (∼1.75 GHz) is observed in the S_{33} curve, which should be due to the mutual coupling between the antenna and the filter. The measured and simulated radiation patterns of the antenna part are illustrated in Fig. 2.33 for both the E- and H-planes, showing reasonable agreement. As expected, the typical bidirectional patterns for the slot antenna [39] are obtained. Figure 2.34 shows the measured antenna gain, which is about 4.5 dBi at resonance.

Ultrawideband Antenna Filter   In this section a three-port antenna filter that is made on a single substrate is proposed for ultrawideband (UWB) applications [11]. It is constructed by closely integrating an UWB slot antenna with an UWB patch filter etched on the opposite side of the same substrate (Fig. 2.35). The coupling between the antenna and the filter is found to be very low, without needing any isolation layers.

The filter part is studied first. Figure 2.36 shows the simulated and measured reflection coefficients and insertion losses of the filter part. It can be observed from the figure that the patch filter has three resonances. The measured 3-dB passband starts from 4 to 7.04 GHz, which agrees reasonably well with the simulated passband (3.93 to 7.1 GHz). The measured insertion loss (S_{21}) is found to be less than 1.5 dB across the entire passband. Next, the antenna part of the IAF is investigated. The simulated and measured VSWRs of the U-shaped slot antenna

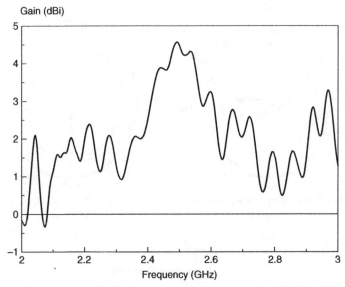

FIGURE 2.34 Measured antenna gain of the antenna filter. (From [9], copyright © 2010 IEEE, with permission.)

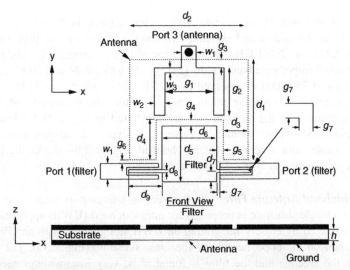

FIGURE 2.35 Antenna filter constructed by integrating an ultrawideband slot antenna and a patch filter: (a) top view; (b) front view. (From [11], copyright © 2011 John Wiley & Sons, Inc., with permission.)

are shown in Fig. 2.37. With reference to the figure, the simulated and measured passbands (VSWR < 2) at port 3 are 3.28 to 10 GHz and 3.17 to 10 GHz, respectively. Figure 2.38 shows the measured and simulated couplings (S_{13}) between the antenna and filter ports. As can be seen from the figure, the coupling is generally

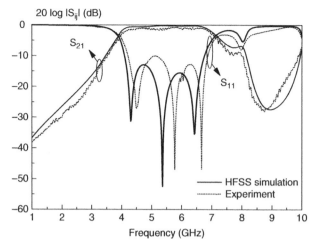

FIGURE 2.36 Simulated and measured reflection coefficients and insertion losses of the filter part shown as a function of frequency. (From [11], copyright © 2011 John Wiley & Sons, Inc., with permission.)

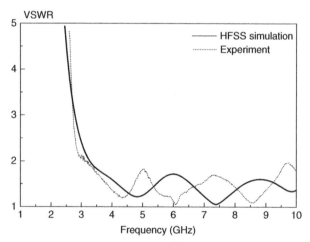

FIGURE 2.37 Simulated and measured VSWRs of the antenna part shown as a function of frequency. (From [11], copyright © 2011 John Wiley & Sons, Inc., with permission.)

weaker than 20 dB, even when $g_4 = g_5 = 0$ was used in the design. The results for S_{23} are similar but are not shown for brevity. Figure 2.39 shows the measured and simulated radiation patterns of the antenna part at 4 and 7 GHz, with reasonable agreement. With reference to the figure, the bidirectional radiations are quite stable as the frequency changes. For all of the cases, the cross-polarized fields are weaker than their co-polarized counterparts by about 20 dB in both the forward ($\theta = 0°$) and backward ($\theta = 180°$) directions. The measured antenna gain is shown in Fig. 2.40, where it is found that the gain varies between 1 and 7 dBi over the frequency range from 3 to 10 GHz.

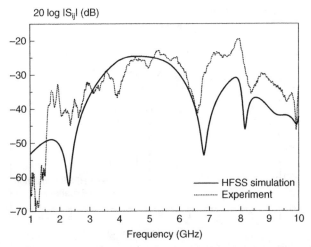

FIGURE 2.38 Simulated and measured coupling coefficients between the antenna and filter parts of the antenna filter in Fig. 2.35. (From [11], copyright © 2011 John Wiley & Sons, Inc., with permission.)

2.3 BALUN FILTERS

A balun (balanced-to-unbalanced) filter is a new component that combines the power dividing and filtering functions into one, with the single- and multiband configurations shown in Fig. 2.41(a) and 2.41(b), respectively. In each Nth of the passbands, a balun filter is designed to distribute an input power (P_{inN}, $N = 1, 2, \ldots$) evenly to a pair of output ports, which have a phase difference of 180° ($|\theta_{2N} - \theta_{3N}| = 180°$). To be a dual function, each of the output ports itself must also be able to be made an independent filter. In recent years, research on the integration of balun and filter using different techniques has been reported , [12–14, 40–44]. A ring resonator [12] and cross-slotted rectangular microstrip patch [40,41] were used for exciting degenerate modes for a dual-mode balun filter. Dual-band cases are discussed in the literature [14,43]. The size of such a multifunctional component can be made very compact by using the LTCC technology [44]. As neither the balun nor filter radiates, a balun filter can usually be modeled with reasonable accuracy by SPICE-based software.

2.3.1 Ring Balun Filter

Configuration Jung and Hwang [12] first deployed a microstrip ring resonator as a dual-functional balun filter. It was made by cascading several sections of transmission lines, each of which has a different impedance, to form a dual-mode ring resonator. The ring has a one-wavelength circumference in total, with the configuration shown in Fig. 2.42. The design parameters are given by $a = 0.60$ mm, $b = 0.75$ mm, $c = 1.46$ mm, $d = 0.30$ mm, $e = 0.90$ mm, $f = 0.36$ mm, $g = 2.40$ mm, $h = 0.10$ mm, $i = 16.14$, and $j = 17.90$ mm [12]. As can be seen from the

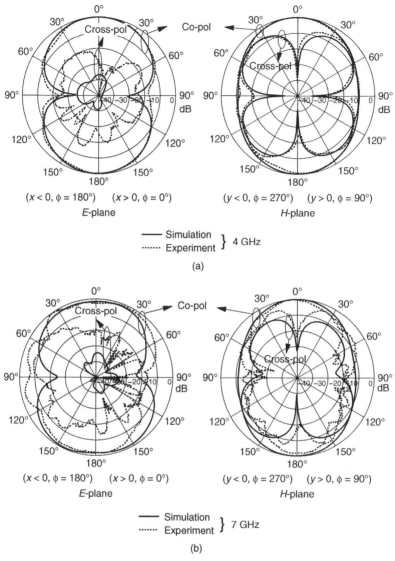

FIGURE 2.39 Simulated and measured normalized radiation patterns of the antenna part of the IAF in Fig. 2.35: (a) 4 GHz; (b) 7 GHz. (From [11], copyright © 2011 John Wiley & Sons, Inc., with permission.)

figure, the resonator is capacitively coupled to an input and to the two outputs via narrow slits. For ease of analysis, it can be represented by a transmission-line model, as shown in Fig. 2.43. With reference to the figure, the ring can be equally divided into four portions, having characteristic impedances of Z_1, Z_2, Z_3, and Z_4. Figure 2.44 shows a prototype balun filter.

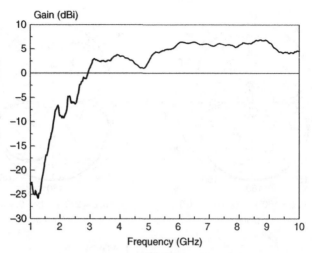

FIGURE 2.40   Measured antenna gain of the antenna part of the antenna filter in Fig. 2.35. (From [11], copyright © 2011 John Wiley & Sons, Inc., with permission.)

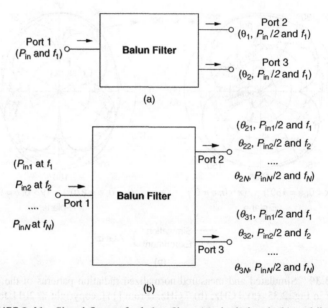

FIGURE 2.41   Signal flows of a balun filter: (a) single-band; (b) multiband.

Results and Discussion   Given in Fig. 2.45 are the simulated S parameters generated by the circuit simulator ADS. Also sketched on the same figure are the results measured. It is observed that the measurement agrees very well with the transmission-line model, given in Fig. 2.43. As can be seen from Fig. 2.45, each of the two output ports has a bandpassing 3-dB bandwidth $[\mathrm{BW}_{3\mathrm{dB}} = (f_H^{-3\mathrm{dB}} - f_L^{-3\mathrm{dB}})/f_0]$ of about 1.6%, operating in the frequency range

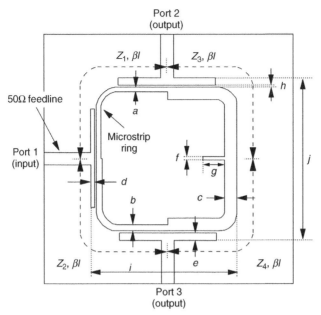

FIGURE 2.42 Top-down view of a microstrip ring balun filter. (Courtesy of H.-Y. Hwang, Kangwon National University. From [12]. copyright © 2007 IEEE, with permission.)

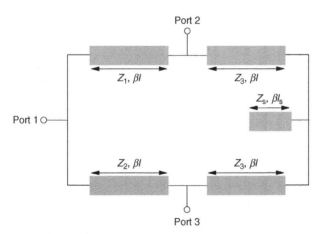

FIGURE 2.43 Equivalent circuit for the microstrip balun filter in Fig. 2.42. (Courtesy of H.-Y. Hwang, Kangwon National University. From [12]. copyright © 2007 IEEE, with permission.)

2.425 to 2.465 GHz. The center frequency is given by $f_0 \sim 2.44$ GHz. $f_H^{-3\text{dB}}$ and $f_L^{-3\text{dB}}$ are the frequencies at which the signal power reduces by half at the upper and lower bounds of the passband, respectively. The insertion loss is about 5.5 dB, as can be seen from Fig. 2.45. Next, the amplitude and phase imbalances of the balun are measured. With reference to Fig. 2.46, the amplitude imbalance

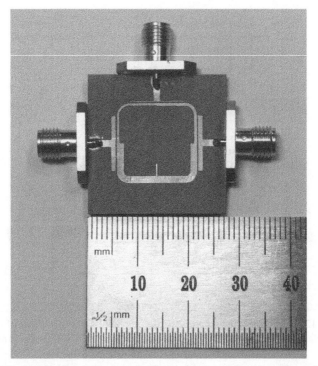

FIGURE 2.44 Balun filter. (Courtesy of H.-Y. Hwang, Kangwon National University. From [12], copyright © 2007 IEEE, with permission.)

and phase difference between the two balanced ports are well within 0.5 dB and $180 \pm 5°$, respectively, in the passband.

2.3.2 Magnetic-Coupled Balun Filter

Configuration   Figure 2.47 shows a balun filter that is designed using a pair of magnetically coupled resonators [13]. As can be seen from the figure, interdigital configuration was used to minimize the circuit size. The design parameters are $l_s = 1.2$ mm, $w_c = 2.5$ mm, $l_c = 1.3$ mm, $g_c = 0.1$ mm, $f_w = 0.2$ mm, and $c_e = 1.7$ mm. It can be represented by the equivalent-circuit model shown in Fig. 2.48, where the component parameters are given by $C = 0.24$ pF, $L = 2.45$ nH, $L_s = 1.2$ nH, $C_g = 0.06$ pF, $C_p = 0.72$ pF, $C_{p'} = 1.24$ pF, $C_e = 0.21$ pF, and $M = 0.08$. The prototype is shown in Fig. 2.49.

Results and Discussion   In Fig. 2.50 an equivalent-circuit model was used to generate the simulation results, then verified with experiments. As can be seen from the figure, the measured reflection coefficient and insertion loss at the unbalanced port are below -15 and -5.5 dB, respectively, in the passband from 2.4 to 2.5 GHz.

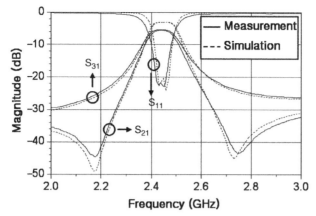

FIGURE 2.45 Simulated and measured S parameters of the microstrip balun filter in Fig. 2.42. (Courtesy of H.-Y. Hwang, Kangwon National University. From [12], copyright © 2007 IEEE, with permission.)

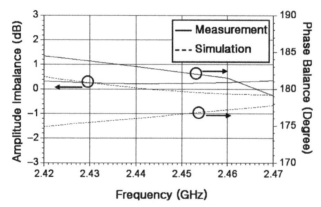

FIGURE 2.46 Simulated and measured amplitude and phase imbalances of the microstrip balun filter in Fig. 2.42. (Courtesy of H.-Y. Hwang, Kangwon National University. From [12], copyright © 2007 IEEE, with permission.)

With reference to Fig. 2.51, the phase difference measured is within $180 \pm 2°$, and the measured amplitude imbalance is well below 2 dB.

2.3.3 Rectangular Patch Balun Filter

Configuration In [45,46], unequal crossed slots were etched on a microstrip rectangular patch for producing double modes in a bandpass filter. This perturbation method is similar to that used by a CP antenna in which two degenerate modes are excited by introducing disturbance to the resonator geometry. The unloaded Q factor of such a filter is usually in the range 100 to 300. With the use of the same technique, a patch balun filter [40] was designed and is shown in Fig. 2.52. As can be seen from the figure, cross-slots are again used for generating two

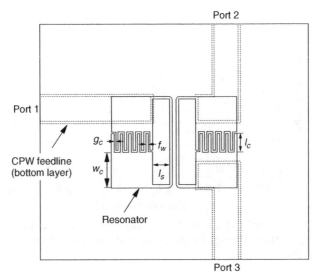

FIGURE 2.47 Top-down view of a CPW magnetic-coupled balun filter. (Courtesy of J. Lee, Chonnam National University. From [13], copyright © 2009 IEEE, with permission.)

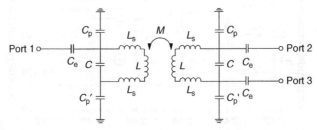

FIGURE 2.48 Equivalent-circuit model of the CPW magnetic-coupled balun filter in Fig. 2.47. (Courtesy of J. Lee, Chonnam National University. From [13], copyright © 2009 IEEE, with permission.)

degenerate modes in a rectangular patch. It works as two independent dual-mode filters in the signal paths of port 1–2 and port 1–3. At the same time, it is a good balun that divides a microwave signal evenly into two with 180° phase difference between them. Long capacitive gaps are placed along the patch edges to effectively couple microwave signals to ports 2 and 3. Other design parameters are given by $W = 18$ mm , $w_1 = 0.933$ mm, $w_2 = 0.2$ mm, $g_1 = 0.2$ mm $g_2 = 0.1335$ mm, $g_3 = 0.1$ mm, $g_4 = 0.07$ mm. $l_1 = 7$ mm, $l_2 = 4$ mm, $d = 2$ mm, and $\phi' = 45°$. It is made on a grounded substrate (50×50 mm^2) with a dielectric constant of $\varepsilon_r = 6.15$ and a thickness of 0.635 mm. The prototype is shown in Fig. 2.53.

Results and Discussion Ansoft HFSS was used to simulate a microstrip patch balun filter, and the results were later verified with experiments. Figure 2.54 shows the simulated and measured S parameters as a function of frequency. In general, reasonable agreement is found between simulation and measurement. For both S_{21}

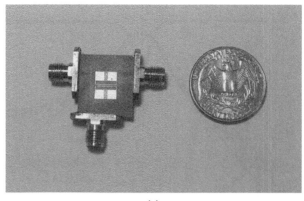

(a)

(b)

FIGURE 2.49 CPW magnetic-coupled balun filter: (a) top view; (b) back view. (Courtesy of J. Lee, Chonnam National University. From [13], copyright © 2009 IEEE, with permission.)

and S_{31}, the measured 3-dB filter passband covers 2.91 to 3.19 GHz, centering at 3.05 GHz. Two transmission zeros are also observed in S_{31} near the filter passband. With reference to the same figure, the output signal magnitudes ($|S_{21}|$ and $|S_{31}|$) are equal at about -6 dB. Additional losses can be introduced by the connectors and feedlines. Figure 2.55 gives the measured output phase of the microstrip patch balun filter. As can be seen from the figure, the phase bandwidth ($180 \pm 5°$) encompasses 2.86 to almost 3.4 GHz.

2.4 ANTENNA PACKAGE

Packaging cover is required by most electronic devices for electrical connection with other circuits, protection from mechanical damage and dirt, dissipation of excess heat generated by the circuit, and shielding from electromagnetic

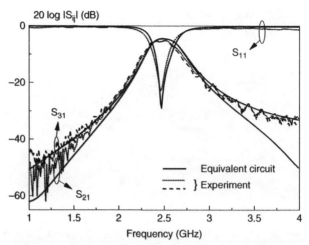

FIGURE 2.50 Measured and modeled S parameters of a magnetic-coupled balun filter in Fig. 2.47. (Courtesy of J. Lee, Chonnam National University. From [13], copyright © 2009 IEEE, with permission.)

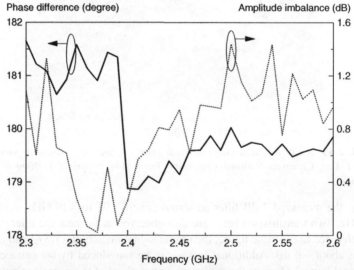

FIGURE 2.51 Measured phase difference and amplitude imbalance of the magnetic-coupled balun filter in Fig. 2.47. (Courtesy of J. Lee, Chonnam National University. From [13], copyright © 2009 IEEE, with permission.)

interference and electrostatic discharge [16]. In modern electronic systems, packages are also designed into different shapes for aesthetics and marketing considerations. As the chemical materials that made the integrated circuits can be harmful to the environment, a package is also required to enhance the product safety and preventing contamination. Figure 2.56 shows some of the contemporary

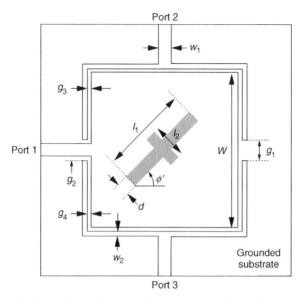

FIGURE 2.52 Top-down view of a microstrip patch balun filter, with crossed slots. (Courtesy of C. H. Ng, Agilent Technologies Sdn. Bhd., Malaysia. From [40].)

FIGURE 2.53 Microstrip patch balun filter. (Courtesy of C. H. Ng, Agilent Technologies Sdn. Bhd., Malaysia. From [40].)

packages for microwave circuits in the commercial market. As the package is much bulkier than a semiconductor chip, there is a strong desire to limit its footprint and size. Recently, some works were conducted to explore using the packaging cover as the radiating element [16,17].

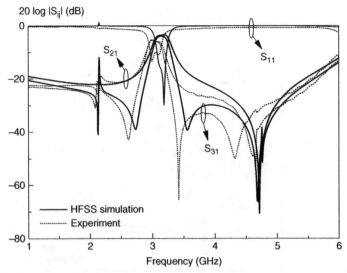

FIGURE 2.54   Simulated and measured S parameters of the microstrip patch balun filter in Fig. 2.52. (Courtesy of C. H. Ng, Agilent Technologies Sdn. Bhd., Malaysia. From [40].)

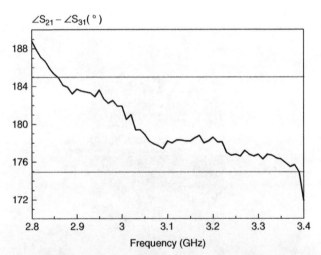

FIGURE 2.55   Measured output phase of the microstrip patch balun filter in Fig. 2.52. (Courtesy of C. H. Ng, Agilent Technologies Sdn. Bhd., Malaysia. From [40].)

2.4.1   DRA Packaging Cover

Configuration   The use of a DRA as a packaging cover was first explored by Lim and Leung [16]. Two configurations were proposed, with the first shown in Fig. 2.57. Being a packaging cover simultaneously, a rectangular hollow DRA (with dimensions $L = 30$ mm, $W = 29$ mm, and $H = 15$ mm) is excited by a microstrip-fed aperture etched on the top of a concentric rectangular metallic cavity (with dimensions $a = 15$ mm, $b = 21.6$ mm, and $h = 5$ mm). In the experiment, a microstrip feedline is connected to the center conductor of an external coaxial line for excitation of the DRA. In practice, however, the feedline can be connected to either active or passive devices that are shielded inside the cavity. The packaging DRA operating at 2.4 GHz can be designed using the following procedure:

1. Use is made of the transcendental equation presented by Mongia and Ittipiboon [47] to obtain the desired dimensions of a rectangular solid DRA for the fundamental TE mode working at 2.4 GHz (or a frequency around 2.4 GHz).

2. The lower center portion of a DRA is removed to form a notched DRA [48]. In this case, the hollow DRA has a higher fundamental resonance frequency than the solid DRA.

3. A hollow DRA is formed by covering up the two sides, parallel to the E-plane, of the notched DRA. The fundamental resonance frequency of

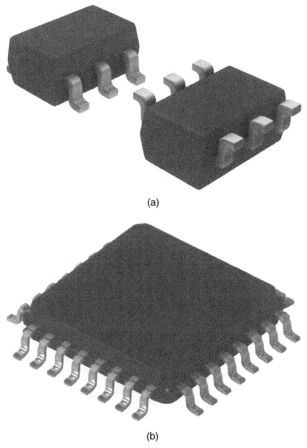

(a)

(b)

FIGURE 2.56 Contemporary microwave packages: (a) small outline package (SOP); (b) quad flat package (QFP); (c) transistor outline (TO) package; (d) dual-flat no-leads (DFN) package.

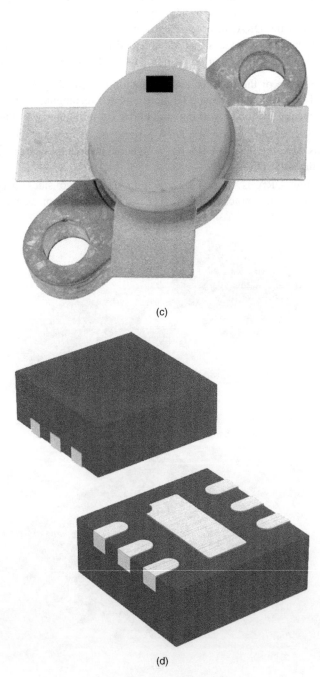

(c)

(d)

FIGURE 2.56   (*Continued*)

the hollow antenna decreases with increasing thickness of the added side-walls. In this DRA configuration, both sidewalls have the same thickness, and the thickness is carefully adjusted to shift the resonant frequency back to 2.4 GHz.

The aforementioned three-step procedure can be used to design and tune the operating frequency of the DRA. The Eccostock HiK Powder [49], a kind of fine-granular and free-flowing powder with dielectric constant of $\varepsilon_r = 12$ was used as the dielectric material. It was noted that slight jogging is generally sufficient to obtain even powder density in the container, and no further postprocessing is required. The dielectric properties of the HiK powder were well tested as described by Junker et al. [50]. In the experiment, a few hard-form clad boards ($\varepsilon_r \sim 1$) were used to construct a rectangular container for the powder. With reference to Fig. 2.57, the faces of the rectangular metallic cavity are formed by a Duroid substrate (top face, with a thickness of 0.762 mm) and four metallic supports (sidewalls), concentrically in the air region of the hollow DRA. A microstrip-fed aperture is etched in the ground plane of the microstrip for exciting the DRA. The aperture has a length of $0.2063\lambda_e$, where $\lambda_e = \lambda_0/\sqrt{(\varepsilon_{rs}+1)/2}$. The ground

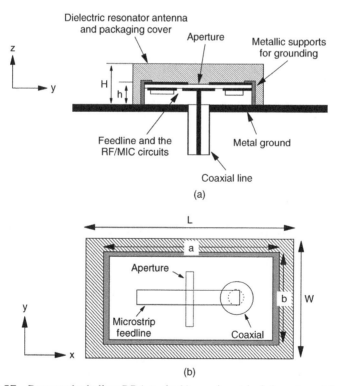

FIGURE 2.57 Rectangular hollow DRA excited by a microstrip-fed aperture: (a) side view; (b) top view. (From [16], copyright © 2006 IEEE, with permission.)

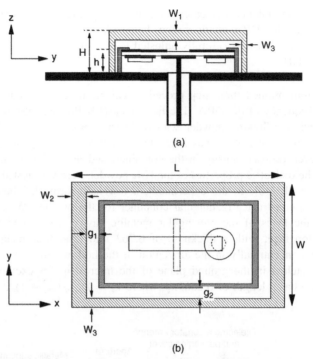

FIGURE 2.58   Rectangular hollow DRA with an airgap, excited by a microstrip-fed aperture: (a) side view; (b) top view. (From [16], copyright © 2006 IEEE, with permission.)

plane of the microstrip also acts as the electrical ground for the RF/MIC circuits located on the opposite side of the substrate, as can be seen in Fig. 2.58. The substrate grounding is connected to the main electrical ground plane through the metallic supports. To demonstrate the design idea, an amplifier is used as the active device.

The second configuration is shown in Fig. 2.58, where the metallic cavity and circuits are inserted into the hollow region of the DRA, which has dimensions of $L = 12.7$ mm, $W = 9$ mm, $H = 6.35$ mm, $h = 3$ mm, $W_1 = 1.85$ mm, $W_2 = 1.85$ mm, $W_3 = 1.3$ mm, $g_1 = 0.055$ mm, and $g_2 = 0.025$ mm. In this case, simulation results show that field coupling can be efficient only when the dielectric constant is sufficiently high. Here, the dielectric constant is chosen to be $\varepsilon_r = 82$. The DRA packaging cover in Fig. 2.58 provides more manufacturing ease, as the dielectric resonator is easily removable. The aforementioned design procedure is applicable to this DRA.

Results and Discussion   In Fig. 2.59 the simulated and measured reflection coefficients are shown as a function of frequency for the antenna in Fig. 2.57. The measured impedance bandwidth ($S_{11} \le -10$ dB) is 5.6%, slightly higher than for the simulation (4%). Good agreement is found between the measured (2.42 GHz) and simulated (2.40 GHz) resonance frequencies. The input impedance is shown

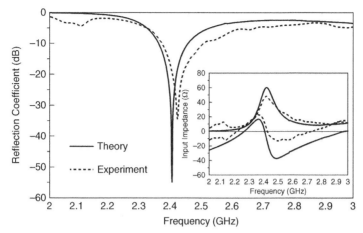

FIGURE 2.59 Simulated and measured reflection coefficients of the hollow DRA with an embedded metallic cavity (Fig. 2.57). The inset shows the simulated and measured input impedances of the hollow DRA as a function of frequency. (From [16], copyright © 2006 IEEE, with permission.)

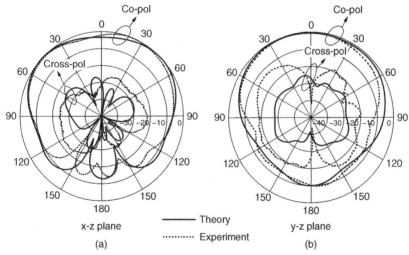

FIGURE 2.60 Simulated and measured normalized radiation patterns of a rectangular hollow DRA in Fig. 2.57: (a) E-plane (xz-plane); (b) H-plane (yz-plane). (From [16], copyright © 2006 IEEE, with permission.)

in the inset of Fig. 2.59. In Fig. 2.60, the simulated and measured normalized radiation patterns are shown for both the E- (xz-plane) and H-plane (yz-plane) at 2.42 GHz, where the broadside TE mode is obtained as expected. The co-polarized fields are in general 20 dB stronger than their cross-polarized counterparts in the boresight direction ($\theta = 0°$).

Figure 2.61 shows the simulated reflection coefficient and normalized radiation patterns for the DRA configuration in Fig. 2.58. The input impedance is also given in the inset of Fig. 2.61(a). With reference to the figure, the simulated impedance bandwidth is about 1.2% at the fundamental resonance frequency of 3.43 GHz. It is reasonable for a high-permittivity DRA. The radiation patterns are broad-side, with the co-polarized fields larger than the cross-polarized ones by at least 20 dB in the bore-sight direction. To further study the integration of the DRA packaging

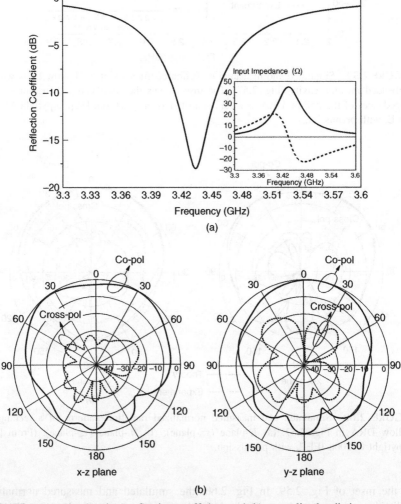

FIGURE 2.61 Simulated (a) reflection coefficient and (b) normalized radiation patterns of the rectangular hollow DRA, with an airgap around the embedded metallic cavity (Fig. 2.58). The inset in part (a) shows the simulated input impedance as a function of frequency. (From [16], copyright © 2006 IEEE, with permission.)

cover with active devices. An Agilent AG302-86 low-noise amplifier (LNA), which has a typical gain of 13.6 dB at 2.4 GHz, was selected to integrate with the DRA shown in Fig. 2.57, whose schematic diagram is shown in the inset of Fig. 2.62. This commercial LNA is prematched to 50 Ω at the input. It is connected to the microstrip feed of the antenna. In an experiment, a small through-hole was drilled on the main ground plane to supply dc bias to the LNA. By combining the DRA and the LNA, an active DRA is designed for transceiver front ends. At the output of the LNA, when it is functioning as a receiver with the dc supply turned on, the measured reflection coefficient of the active DRA is shown in Fig. 2.62. It is noted that the parasitics of the amplifier have shifted the fundamental resonance frequency to around 2.5 GHz, slightly higher than that for the passive DRA. With reference to the figure, two dips are observed at 2.58 and 2.66 GHz, which can be caused by the matching circuits. To investigate the effect of the amplifier on the transmitting characteristics, the active DRA is reconfigured as a transmitter using the same amplifier, and the measured reflection coefficient is also shown in Fig. 2.62. Again, the resonance frequency has drifted to about 2.5 GHz.

Figure 2.63 shows the measured co-polarized radiation pattern of the active DRA (in receiver mode) in the E-plane (the xz-plane) at 2.42 GHz. An amplifier gain ranging from 7 to 12 dB is observed across the observation angle for the active DRA. The cross-polarized fields of both the passive and active DRAs are also plotted in Fig. 2.63. For ease of comparison, all of the radiation patterns are normalized to the maximum value of the co-polarized field of the active DRA. The gain loss (as compared to specification data) can be caused by unavoidable impedance variations and imperfections in the experiment.

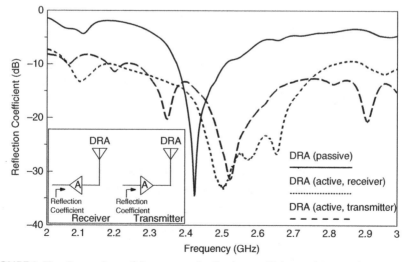

FIGURE 2.62 Comparison of the measured reflection coefficients of the passive and active DRAs (as receiver or transmitter) in Fig. 2.57. (From [16], copyright © 2006 IEEE, with permission.)

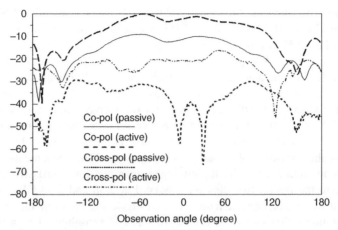

FIGURE 2.63 Comparison of measured E-plane co-polarized and cross-polarized fields between the passive and active DRAs in Fig. 2.57. (From [16], copyright © 2006 IEEE, with permission.)

2.4.2 Other Antenna Packages

In Fig. 2.64 a cylindrical DRA was stacked on top of a shielded package for use as a USB dongle. The DRA was excited by a meandered magnetic loop, being fed by a co-planar waveguide. Other microwave components and devices are all encapsulated inside a shielded metallic cavity. Here, the DR is used simultaneously as a radiating element and a protective layer for the underlying cavity. As the DRA occupies virtually no extra space, the dongle can achieve a very compact size. Such a top-loading DRA can easily be made on a chip to operate at millimeter-wave frequencies by employing the semiconductor fabrication technology [51].

Recently, antenna-on-package (AOP) [52,53], antenna-in-package (AiP) [54], and low-temperature co-fired ceramics (LTCC) [55] techniques have been applied to integrate an antenna with various semiconductor chips and packages. The typical configurations are shown in Fig. 2.65. As can be seen from Fig. 2.65(a), the AOP technology stacks an antenna directly on top of a packaging cover using the

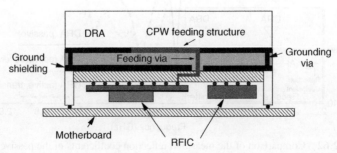

FIGURE 2.64 Side view of a wireless module proposed by Rotaru et al. [17] using the package integrated antenna approach.

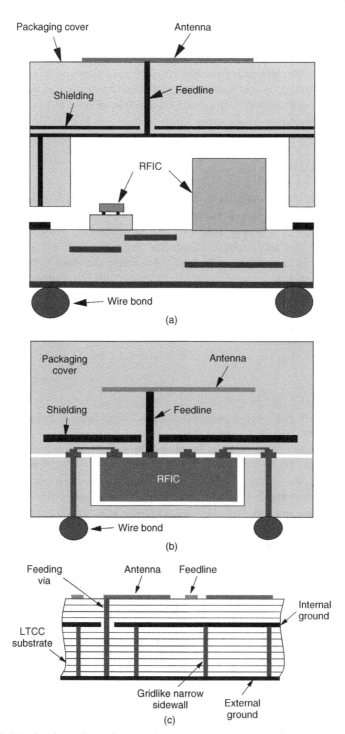

FIGURE 2.65   Configurations of various integrated antennas: (a) AOP (from [52]); (b) AiP (from [54]); (c) LTCC (from [55]).

system-on-package (SOP) technology [37,53]. On the other hand, an antenna is usually embedded inside the package for the AiP [54], as can be seen from Fig. 2.65(b). With the recent developments in material science, the LTCC technique has become a popular way for fabricating microwave- and millimeter-wave systems [55,56]. The main advantage is that densely packed passive circuits can be squeezed into multiple layers of stacked dielectrics. An example of an LTCC millimeter-wave system that is made of 12 dielectric layers is shown in Fig. 2.65(c).

2.5 CONCLUSIONS

In this chapter, several multifunction microwave modules, such as an antenna filter, a balun filter, and an antenna packaging cover, have been discussed. The art and science of PICs is still in a state of rapid development. Market demand for a more compact and less expensive multifunction component is constantly on the rise. As metal is lossy in the millimeter-wave range, the trend is towards merging several functions into a single module for a shorter signal transmission path. Better signal integrity is usually achievable by having a smaller component number and shortening the connection wires on a circuit board. It is foreseeable that in the near future more and more multifunction microwave- and millimeter-wave devices will be explored and studied.

REFERENCES

[1] J. Lin and T. Itoh, "Active integrated antennas," *IEEE Trans. Microwave Theory Tech.*, vol. 42, pp. 2186–2194, Dec. 1994.

[2] Y. Qian and T. Itoh, "Progress in active integrated antennas and their applications," *IEEE Trans. Microwave Theory Tech.*, vol. 46, pp. 1891–1900, Nov. 1998.

[3] K. Chang, R. A. York, P. S. Hall, and T. Itoh, "Active integrated antennas," *IEEE Trans. Microwave Theory Tech.*, vol. 50, pp. 937–944, Mar. 2002.

[4] K. C. Gupta and P. S. Hall, *Analysis and Design of Integrated Circuit Antenna Modules*. New York: Wiley, 2000.

[5] J. A. Navarro and K. Chang, *Integrated Active Antennas and Spatial Power Combining*. New York: Wiley, 1996.

[6] E. H. Lim and K. W. Leung, "Use of the dielectric resonator antenna as a filter element," *IEEE Trans. Antennas Propag.*, vol. 56, pp. 5–10, Jan. 2008.

[7] L. K. Hady, D. Kajfez, and A. A. Kishk, "Triple mode use of a single dielectric resonator," *IEEE Trans. Antennas Propag.*, vol. 57, pp. 1328–1335, May 2009.

[8] L. K. Hady, D. Kajfez, and A. A. Kishk, "Dielectric resonator antenna in a polarization filtering cavity for dual function applications," *IEEE Trans. Microwave Theory Tech.*, vol. 56, pp. 3079–3085, Dec. 2008.

[9] E. H. Lim and K. W. Leung, "Compact three-port integrated-antenna–filter modules," *IEEE Antennas Wireless Propag. Lett.*, vol. 9, pp. 467–470, May 2010.

[10] Y. J. Sung, "Microstrip resonator doubling as a filter and as an antenna," *IEEE Antennas Wireless Propag. Lett.*, vol. 8, pp. 486–489, Mar. 2009.

[11] E. H. Lim and K. W. Leung, "Ultrawideband microstrip integrated-antenna–filter," *Microwave Opt. Tech. Lett.*, vol. 53, no. 1, pp. 32–34, Jan. 2011.

[12] E. Jung and H. Hwang, "A balun-BPF using a dual mode ring resonator," *IEEE Microwave Wireless Compor. Lett.*, vol. 17, pp. 652–654, Sept. 2007.

[13] J. Lee, H. Lee, H. Nam, and Y. Lim, "Balun-BPF using magnetically coupled interdigital-loop resonators," 6th International Conference on Electrical Engineering/Electronics, Computer, Telecommunications and Information Technology, 2009, Pattaya, May 2009, pp. 932–935.

[14] L. K. Yeung and K. L. Wu, "A dual-band coupled-line balun filter," *IEEE Trans. Microwave Theory Tech.*, vol. 55, pp. 2406–2411, Nov. 2007.

[15] Z. Zuo, X. Chen, G. Han, L. Li, and W. Zhang, "An integrated approach to RF antenna–filter co-design," *IEEE Antennas Wireless Propag. Lett.*, vol. 8, pp. 141–144, Apr. 2009.

[16] E. H. Lim and K. W. Leung, "Novel application of the hollow dielectric antenna as a packaging cover," *IEEE Trans. Antennas Propag.*, vol. 54, pp. 484–487, Feb. 2006.

[17] M. Rotaru, Y. Y. Lim, H. Kuruveettil, R. Yang, A. P. Popov, and C. Chua, "Implementation of packaged integrated antenna with embedded front end for Bluetooth applications," *IEEE Trans. Adv. Packag.*, vol. 31, no. 3, pp. 558–567, Aug. 2008.

[18] R. S. Adams, B. O. Neil, and J. F. Young, "The circulator and antenna as a single integrated system," *IEEE Antennas Wireless Propag. Lett.*, vol. 8, pp. 165–168, May 2008.

[19] W. Lim, W. Son, K. S. Oh, W. Kim, and J. Yu, "Compact integrated antenna with circulator for UHF RFID system," *IEEE Antennas Wireless Propag. Lett.*, vol. 7, pp. 673–675, Mar. 2008.

[20] S. B. Cohn, "Microwave bandpass filters containing high-Q dielectric resonators," *IEEE Trans. Microwave Theory Tech.*, vol. 16, pp. 218–227, Apr. 1968.

[21] H. Abe, Y. Takayama, A. Higashisaka, and H. Takamizawa, "A highly stabilized low-noise GaAs FET integrated oscillator with a dielectric resonator in the C band," *IEEE Trans. Microwave Theory Tech.*, vol. 26, pp. 156–162, Mar. 1978.

[22] S. A. Long, M. W. McAllister, and L. C. Shen, "The resonant cylindrical dielectric cavity antenna," *IEEE Trans. Antennas Propag.*, vol. 31, pp. 406–412, May 1983.

[23] K. M. Luk and K. W. Leung, Eds., *Dielectric Resonator Antennas*. Hertfordshire, UK: Research Studies Press, 2003.

[24] A. Petosa, *Dielectric Resonator Antenna Handbook*. Norwood, MA: Artech House, 2007.

[25] D. Kajfez and P. Guillon, *Dielectric Resonators*. Atlanta, GA: Noble, 1998.

[26] T. D. Iveland, "Dielectric resonator filters for application in microwave integrated circuits," *IEEE Trans. Microwave Theory Tech.*, vol. 19, pp. 643–652, July 1971.

[27] C. Wang, K. A. Zaki, A. E. Atia, and T. G. Dolan, "Dielectric combline resonators and filters," *IEEE Trans. Microwave Theory Tech.*, vol. 46, pp. 2501–2506, Dec. 1998.

[28] S. J. Fiedziuszko and S. Holme, "Dielectric resonators raising your high-Q," *IEEE Microwave Mag.*, Sept. 2001.

[29] I. J. Bahl and P. Bhartia, *Microstrip Antennas*. Norwood, MA: Artech House, 1980.

[30] J. S. Hong and M. J. Lancaster, *Microstrip Filters for RF/Microwave Applications*. New York: Wiley, 2001.

[31] L. H. Hsieh and K. Chang, "Equivalent lumped elements G, L, C, and unloaded Q's of closed- and open-loop ring resonators," *IEEE Trans. Microwave Theory Tech.*, vol. 50, pp. 453–460, Feb. 2002.

[32] R. R. Mansour, "Filter technologies for wireless base stations," *IEEE Antennas Propag. Mag.*, vol. 5, no. 1, pp. 68–74, Mar. 2004.

[33] E. Pistono, P. Ferrari, L. Duvillaret, J. Duchamp, and R. G. Harrison, "Hybrid narrow-band tunable bandpass filter based on varactor loaded electromagnetic-bandgap coplanar waveguides," *IEEE Trans. Microwave Theory Tech.*, vol. 53, pp. 2506–2514, Aug. 2005.

[34] S. K. Vaibhav, V. V. Mishra, and A. Biswas, "A modified ring dielectric resonator with improved mode separation and its tunability characteristics in MIC environment," *IEEE Trans. Microwave Theory Tech.*, vol. 53, pp. 1960–1967, June 2005.

[35] Y. P. Zhang, X. J. Li, and T. Y. Phang, "A study of dual-mode bandpass filter integrated in BGA package for single-chip RF transceivers," *IEEE Trans. Adv. Packag.*, vol. 29, no. 2, pp. 354–358, May 2006.

[36] L. K. Yeung and K. L. Wu, "An LTCC balanced-to-unbalanced extracted-pole bandpass filter with complex load," *IEEE Trans. Microwave Theory Tech.*, vol. 54, pp. 1512–1518, Apr. 2006.

[37] Y. P. Zhang, "Integrated circuit ceramic ball grid array package antenna," *IEEE Trans. Antennas Propag.*, vol. 52, pp. 2538–2544, Oct. 2004.

[38] K. Chang and L. H. Hsieh, *Microwave Ring Circuits and Related Structures*. Hoboken, NJ: Wiley, 2004.

[39] R. Garg, P. Bhartia, I. Bahl, and A. Ittipiboon, *Microstrip Antenna Design Handbook*. Dedham, MA: Artech House, 2001.

[40] C. H. Ng, E. H. Lim, and K. W. Leung, "Microstrip patch balun filter," *Cross Strait Quad-Regional Radio Science and Wireless Technology Conference*, Harbin, China, July 28, 2011.

[41] S. Sun and W. Menzel, "Novel dual-mode balun bandpass filters using single cross-slotted patch resonator," *IEEE Microwave Wireless Compon. Lett.*, vol. 21, pp. 415–417, Aug. 2011.

[42] C. H. Hu, C. H. Wang, S. Y. Chen, and C. H. Chen, "Balanced-to-unbalanced bandpass filters and the antenna application," *IEEE Trans. Microwave Theory Tech.*, vol. 56, pp. 2474–2482, Nov. 2008.

[43] G. S. Huang and C. H. Chen, "Dual-band balun bandpass filter with hybrid structure," *IEEE Microwave Wireless Compon. Lett.*, vol. 21, pp. 356–358, July 2011.

[44] G. H. Huang, C. H. Hu, and C. H. Chen, "LTCC balun bandpass filters using dual-response resonators," *IEEE Microwave Wireless Compon. Lett.*, vol. 21, pp. 483–485, Sept. 2011.

[45] L. Zhu, P. M. Wecowski, and K. Wu, "New planar dual-mode filter using cross-slotted patch resonator for simultaneous size and loss reduction," *IEEE Trans. Microwave Theory Tech.*, vol. 47, no. 5, pp. 650–654, May 1999.

[46] L. H. Hsieh and K. Chang, "Compact size and low insertion loss Chebyshev-function bandpass filters using dual-mode patch resonators," *Electron. Lett.*, vol. 37, no. 17, Aug. 2001.

[47] R. K. Mongia and A. Ittipiboon, "Theoretical and experimental investigations on rectangular dielectric resonator antennas," *IEEE Trans. Antennas Propag.*, vol. 45, pp. 1348–1356, Sept. 1997.

[48] A. Petosa, A. Ittipiboon, Y. M. M. Antar, D. Roscoe, and M. Cuhaci, "Recent advances in dielectric-resonator antenna technology," *IEEE Antennas Propag. Mag.*, vol. 40, pp. 35–48, June 1998.

[49] Product Manual for Eccostock *HiK* Powder, Emerson & Cuming Microwave Product. http://www.eccosorb.com/catalog/eccostock/HIKPOWDER.pdf.

[50] G. P. Junker, A. A. Kishk, and A. W. Glisson, "Input impedance of dielectric resonator antennas excited by a coaxial probe," *IEEE Trans. Antennas Propag.*, vol. 42, pp. 960–966, July 1994.

[51] M. Nezhad-Ahmadi, M. Fakharzadeh, B. Biglarbegian, and S. Safavi-Naeini, "Input impedance of dielectric resonator antennas excited by a coaxial high-efficiency on-chip dielectric resonator antenna for mm-wave transceivers," *IEEE Trans. Antennas Propag.*, vol. 58, pp. 3388–3392, Oct. 2010.

[52] S. Brebels, J. Ryckaert, B. Come, S. Donnay, W. D. Raedt, E. Beyne, and R. P. Mertens, "SOP integration and codesign of antennas," *IEEE Trans. Adv. Packag.*, vol. 27, no. 2, pp. 341–351, May 2004.

[53] M. M. Tentzeris, J. Laskar, J. Papapolymerou, S. Pinel, V. Palazzari, G. DeJean, N. Papageorgiou, J. Thompson, R. Bairavasubramanian, S. Sarkar, and J. H. Lee, "3-D-integrated RF and millimeter-wave functions and modules using liquid crystal polymer (LCP) system-on-package technology," *IEEE Trans. Adv. Packag.*, vol. 27, no. 2, pp. 332–340, May 2004.

[54] Y. P. Zhang, "Antenna-on-chip and antenna-in-package solutions to highly integrated millimeter-wave devices for wireless communications," *IEEE Trans. Antennas Propag.*, vol. 52, pp. 2538–2544, Oct. 2004.

[55] Y. Huang, K. L. Wu, D. G. Fang, and M. Ehlert, "An integrated LTCC millimeter-wave planar array antenna with low-loss feeding network," *IEEE Trans. Antennas Propag.*, vol. 53, pp. 1232–1234, Mar. 2005.

[56] Y. Q. Zhang, Y. X. Guo, and M. S. Leong, "A novel multilayer UWB antenna on LTCC," *IEEE Trans. Antennas Propag.*, vol. 58, pp. 3013–3019, Sept. 2010.

CHAPTER THREE

Reconfigurable Antennas

XUE-SONG YANG, SHAO-QIU XIAO, and BING-ZHONG WANG

3.1 INTRODUCTION

To support a variety of applications, such as communication, navigation, and surveillance, most wireless systems require more than one antenna. These antennas, which may work at different frequencies and polarizations, are usually installed at a number of positions on a wireless platform, such as a radar station, a satellite base station, or a mobile phone, for better reception quality. The use of multiple antennas is definitely very undesirable, as it can increase system size and material costs. Worse still, the antennas may introduce electromagnetic interference, which can jeopardize normal operation of the electronic circuits. It is obvious that a possible way to cut down the number of antennas is to have one that can be reconfigured to provide several functions and can also operate at different frequencies, switch field polarizations, and sweep the radiation beams. Such a new multifunction antenna configuration is called a *reconfigurable antenna*. The concept of a reconfigurable antenna first appeared in a U.S. patent in 1983 [1]. Since then it has been well received by academics, industry, and the military. In general, an antenna can easily be made reconfigurable by either incorporating active devices or by having multiple input ports. For the first type, switches such as microelectromechanical system (MEMS) switches, PIN diode switches, and optical switches are usually employed to alter the radiation aperture of an antenna. In the latter, additional feeding ports are used to tap out the different operational modes of an antenna, mainly to obtain different polarizations and radiation patterns.

Compact Multifunctional Antennas for Wireless Systems, First Edition. Eng Hock Lim, Kwok Wa Leung.
© 2012 John Wiley & Sons, Inc. Published 2012 by John Wiley & Sons, Inc.

Reconfigurable antennas can be classified according to their functions, operating frequencies, geometries, materials, and so on. Among them, classification based on functionalities is the most popular. An antenna that has a variable frequency is called a *frequency-reconfigurable antenna*, one that has a changeable radiation pattern is called a *pattern-reconfigurable antenna*, and one with switchable polarization is called a *polarization-reconfigurable antenna*. A reconfigurable antenna that includes multiple functions is termed a *multi-reconfigurable antenna*.

Frequency and pattern reconfigurations can be accomplished in many ways, such as changing the resonance length, re-routing the aperture current, incorporating variable loads to change the current behavior of the antenna, adding variable parasitic elements to alter the performance of the driven element, or combining different types of antennas together using switches. In Section 3.2, the design considerations of reconfigurable antennas are described. Sections 3.3 and 3.4 provide case studies on the frequency- and pattern-reconfigurable antennas. Several polarization-reconfigurable antennas are explored in Section 3.5. Finally, some multi-reconfigurable antennas, being reconfigurable in both frequency and radiation pattern, are introduced in Section 3.6.

3.2   DESIGN CONSIDERATIONS AND RECENT DEVELOPMENTS

In this section, designs of planar-reconfigurable antennas are discussed. In the early days it was difficult to obtain high-end semiconductor devices commercially, so many reconfigurable antennas were made using mechanical switches, referred to more commonly as *ideal switches*. Such switches are ideal and have low loss, but they do not allow fast switching speed and automation. Today, with the rapid development of MEMS, semiconductor, and photoconductor technologies, most reconfigurable antennas can be fabricated easily using different diodes. A PIN diode, which can easily be made into a simple switch, has been explored extensively for designing various reconfigurable antennas because of a number of advantages, including good reliability, compact size, high switching speed, and low resistance and capacitance in the ON and OFF states. An operating PIN diode requires a dc bias, and its equivalent circuits are shown in Fig. 3.1 for both the ON (forward-biased) and OFF (reverse-biased) states. When turned on, a diode can be represented by a forward-biased resistance R_S (usually, ~ 0.1 Ω at 1 mA) and a package inductance L (~ 0.1 nH). As can be seen in the figure, an OFF diode is represented by a very large reverse resistance R_P and a package capacitance C_T. To simplify the circuit analysis, the ON and OFF states of a diode are simply treated as ideal switches, as shown in Fig. 3.2. In the past decade, the PIN diode has been utilized in designing various frequency- and pattern-reconfigurable antennas [2–4]. Although the semiconductor-made switch enables fast configuration speed, it is lossy and can seriously affect the antenna performance if not designed properly.

Peroulis et al. used a PIN diode to configure the operating frequency of a slot antenna [3]. A transmission-line equivalent circuit was proposed to study the

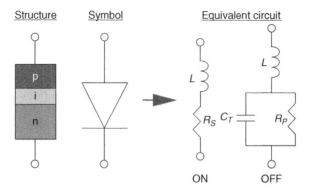

FIGURE 3.1 Equivalent circuits of a PIN diode at the ON and OFF states.

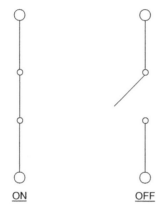

FIGURE 3.2 Ideal equivalent circuits of a PIN diode at the ON state (forward biased) and at the OFF state (reverse biased).

effect of the diode on a frequency-reconfigurable antenna. By controlling its biasing voltage, a diode can easily be turned ON and OFF. It is always very desirable to have a diode that has a low loading effect (OFF state) and low resistance (ON state). Obviously, the ON-state resistance can result in power dissipation and degrade the antenna radiation efficiency. For such a reconfigurable antenna, the total dissipated power depends on the diode's ON resistance and the number of switches being used, and these are inherent drawbacks. Analysis shows that even a small series resistance of $R = 1.4 \ \Omega$ can cause the antenna gain to deteriorate (2.5 dB lower than the ideal value). In simulations, a lumped resistor can be used to model the loading effect and resistance of the diode. The PIN diode can also cause the resonance frequency of the slot antenna to fluctuate, especially when more than one is employed for multifrequency operation. For the reconfigurable antenna reported in [3], it was found that the reflection coefficient is less than -20 dB when the diode is in the OFF state. For both the ON and OFF states, the RF–dc isolation can be made better than 30 dB, with the RF–RF isolation always greater than 10 dB up to 1 GHz. The use of high-frequency diodes or RF MEMS switches can further increase the isolation for applications beyond UHF.

The performance of a reconfigurable PIN-driven slot-loaded patch antenna was compared experimentally with one made with ideal switches [4]. It was found that the antenna loaded with an ON diode had a lower resonance frequency than that using an ideal switch. The reverse was seen for the antenna in the OFF state. It is worth noting that the reconfigurable antenna has very similar radiation patterns in both states. This simply implies that the diode has nearly no effect on the radiation characteristics of the antenna. A 1-dB loss of antenna gain was observed when the diode was turned on. Similar results were found by Kang et al. [2].

As the quality factor (Q) of the PIN diode is usually low at high frequencies ($Q < 3$ at 10 GHz), it has a high insertion loss and can reduce the antenna gain of a reconfigurable antenna. The use of MEMS switches, which have relatively higher Q values ($Q > 10$ at 10 GHz) and lower loss, can help alleviate this problem [5]. Also, most MEMS devices come with a very wide operating frequency range and low power consumption. They can also be integrated monolithically with antennas and other semiconductor components. Two types of MEMS switches are shown in Fig. 3.3. Although there are still some problems, such as switching reliability and speed, the use of MEMS switches has become very common, and it is now the most feasible alternative [6] in designs of reconfigurable antennas. A single-substrate MEMS-integrated reconfigurable antenna was proposed by Jung et al. [6] to provide beam steering capability.

Different active devices, such as MEMS, PIN diode, and FET switches, were used by Mak et al. [7] to switch the feed port and ground of a planar inverted F antenna (PIFA). The reconfiguration performances were compared. It was found that all three active devices had a reactive loading effect on the antenna. It caused the resonance frequency of the PIFA to decrease. Also, the PIN diode was found to have the least reactive loading. All three active devices degraded the radiation efficiency of the PIFA.

Photoconducting switches were used by Panagamuwa et al. [8] to reconfigure a dipole. When it was turned off, the switch had a weak conductive effect on the antenna, causing the resonance frequency to decrease. As the photoconducting switch is transparent to electromagnetic waves, it does not disturb the antenna radiation.

Recent developments for frequency-, pattern-, and multiple-reconfigurable antennas are discussed next. Some design examples are also given.

3.3 FREQUENCY-RECONFIGURABLE ANTENNAS

Frequency reconfiguration is usually accomplished by changing the operating frequency of an antenna while keeping its radiation characteristics unchanged. The operating frequency of a resonant antenna is changeable by incorporating some switches to alter the resonance length or aperture of its resonator; alternatively, it can be done by switching the feeding ports or antenna grounds [7]. In past decades,

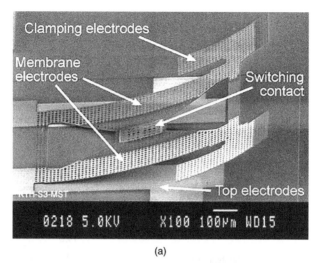

(a)

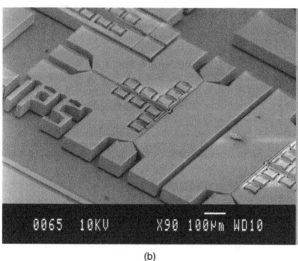

(b)

FIGURE 3.3 Two types of MEMS switches: (a) electrostatic S-shaped film switch; (b) multi-stable cantilever switch. (From http://www.ee.kth.se/php/index.php?action=research&cmd=showproject&id=23.)

many studies on frequency reconfiguration have been conducted on the slot antenna [3,9], microstrip Yagi antenna [10], slit-loaded patch antenna [11], and slot-loaded E-shaped patch [12]. Frequency-reconfigurable antennas are utilized widely in multifrequency and frequency-agile wireless systems.

A frequency-reconfigurable S-shaped slot antenna loaded with a series of PIN diode switches was studied by Peroulis et al. [3]. The configuration of the antenna is shown in Fig. 3.4. Tuning the antenna operating frequency can be done easily by changing its effective electrical length, which is controlled by the bias voltages

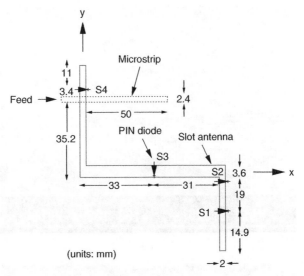

FIGURE 3.4 Configuration of a reconfigurable slot antenna. (From [3], copyright © 2005 IEEE, with permission.)

of solid-state shunt PIN diode switches placed along the slot resonator. Here, four switches are incorporated into the slot to tune the antenna over the frequency range 540 to 950 MHz. When any of the switches is set to the ON state, the slot is shorted at that location, resulting in a decrease in the slot length and an increase in the resonance frequency. Simulations were conducted using the method of moments. The reconfigurable antenna (Fig. 3.4) was fabricated on a 100-mil-thick RT/Duroid substrate with a ground plane of 5×5 in^2. The measured resonances of the antenna are shown in Table 3.1, together with the switch settings. Satisfactory agreement was obtained between the theoretical and experimental data. However, for the highest resonance frequency, there is a discrepancy of 13% between the transmission-line

TABLE 3.1 Comparison of Measured and Calculated Resonance Frequencies

Simulation			Measurement	
TLM[a] f_R (MHz)	MoM[b] f_R (MHz)	Switch Configuration	f_R (MHz)	Switch Configuration
542	561	4 = ON; 1, 2, 3 = OFF	537	4 = ON; 1, 2, 3 = OFF
596	627	1, 4 = ON; 2, 3 = OFF	603	1, 4 = ON; 2, 3 = OFF
688	711	2, 4 = ON; 1, 3 = OFF	684	2, 4 = ON; 1, 3 = OFF
1002	950	3 = ON; 1, 2, 4 = OFF	887	2, 3 = ON; 1, 4 = OFF

Source: [3].
[a] TLM, transmission-line model.
[b] MoM, moment method.

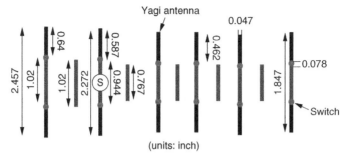

FIGURE 3.5   Dual-band reconfigurable Yagi antenna. (From [10], copyright © 2003 John Wiley & Sons, Inc., with permission.)

model and the measurement, and this can be attributed to the inadequacy of the equivalent circuit (for the diode) at high frequencies.

A dual-band frequency-reconfigurable Yagi antenna was studied by Wahid et al. [10]. The geometry of a reconfigurable Yagi antenna is shown in Fig. 3.5. It consists of two independent Yagi antennas which are made on the same plane, switching between 2.4 and 5.78 GHz. The Yagi antenna works at 2.4 GHz when all the switches are in the ON state. The same antenna operates at 5.78 GHz when all the switches are OFF. Commercial software IE3D was used to simulate the reconfigurable Yagi antenna. With the use of ideal switches, the antenna is simulated at the ON [Fig. 3.6(a)] and OFF [Fig. 3.6(b)] states. With reference to Fig. 3.6(a), the Yagi elements have a longer electrical length when all the switches are turned on. This causes the antenna to work at 2.4 GHz. When all the switches are in the OFF state, as shown in Fig. 3.6(b), the electrical length is then truncated, causing the antenna to work at about 5.8 GHz. Without including the switches, the two antennas (Fig. 3.6) were made. The dimensions of the optimized Yagi antennas are given in Table 3.2. An RT/Duroid 5880 substrate with a thickness of 5 mils, a dielectric permittivity ε_r of 2.2, and a loss tangent of 0.0009 was used for the experiments. The effect of the substrate can be omitted because of its small thickness and low loss. The measured and simulated results are compared in Table 3.3, with good agreement observed. The discrepancy between the simulated and measured results can be caused by the dielectric and conductor losses.

3.3.1   Frequency-Reconfigurable Slot-Loaded Microstrip Patch Antenna

Configuration   A reconfigurable microstrip-fed patch antenna was proposed for multifrequency applications by Xiao et al. [11]. The configuration of the antenna is shown in Fig. 3.7. As can be seen from the figure, multiple slits are cut along the nonradiating edges of a rectangular microstrip patch to increase the electric current route on the resonator. It causes the resonance frequency to be reduced. Two pairs of short slits (slits 1–1' and 2–2') and three pairs of long slits (slits 3–3', 4–4', and 5–5') are inserted into the two nonradiating edges of the patch. The short slits 6–6' are used to control the impedance matching of the feedline. To have a higher

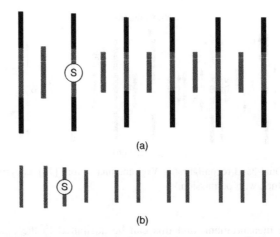

FIGURE 3.6 Configuration of a reconfigurable Yagi antenna: (a) ON state; (b) OFF state.

TABLE 3.2 Dimensions (Inches) of Yagi Antennas Optimized at 2.4 and 5.78 GHz

	Yagi Antenna at 2.4 GHz	Yagi Antenna at 5.78 GHz
Length of driven element (0.42λ)	2.272	0.944
Length of reflector (0.499λ)	2.457	1.02
Length of directors (0.375λ)	1.847	0.767
Spacing between driven element and reflector (0.25λ)	1.23	0.51
Spacing between directors (0.3λ)	1.48	0.61

TABLE 3.3 Comparison of Simulated and Measured Data

	Parameter	Simulation	Measurement
Yagi antenna at 2.4 GHz	3-dB beamwidth	57.6°	50.5°
	Directivity	8.62 dB	7.6 dB
	Front-to-back ratio	13 dB	16.6 dB
Yagi antenna at 5.78 GHz	3-dB beamwidth	37.4°	31.0°
	Directivity	11.4 dB	9.8 dB
	Front-to-back ratio	12 dB	12.5 dB

level of controllability, multiple switches are incorporated uniformly into the slits. With reference to the figure, six and eight switches are used for the short and long slits, respectively. The two slits in each pair (1-1', 2-2', 3-3', 4-4', 5-5', and 6-6') are configured into the same state. The operating frequency of the antenna can easily be shifted by controlling the switches. For each operating frequency,

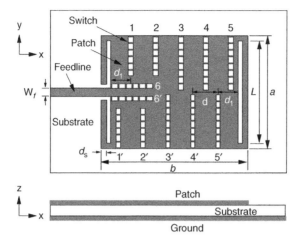

FIGURE 3.7 Configuration of a frequency-reconfigurable slot-loaded microstrip patch antenna. (From [11], copyright © 2003 John Wiley & Sons, Inc., with permission.)

good impedance matching can be obtained by manipulating the switches along the feedline (slit 6–6′).

Results and Discussion   The simulated frequencies of a reconfigurable antenna are listed in Table 3.4. In each slit, switches are numbered from 1 to n, starting from the edge to the middle of the patch. The symbol M_n denotes a switch state in which switches from 1 to n are open (with others closed) in the Mth pair of slits. For example, M_0 is a state where all the switches in the Mth pair of slits are closed. From the table it is clear that the reconfigurable antenna can shift its operating frequency over a one-octave bandwidth from 0.6 to 1.2 GHz by tuning the switches. It is worth noting that the switch settings shown in Table 3.4 are not unique. In other words, there is always more than one switch combinations that can make the antenna operate at a certain frequency. Measurements conducted by using the ideal switches are also given for some of the states. The measured and simulated results agree with each other very well. Since the operation mode for all the states in Table 3.4 is the TM_{10} mode, the radiating patterns are essentially unchanged for all the antenna states. Due to slit loading, the dimension of the radiating patch is decreased about 46% at 0.63 GHz compared with that of the conventional patch antenna.

3.3.2 Frequency-Reconfigurable E-Shaped Patch Antenna

Configuration   An E-shaped microstrip patch antenna, which is fed by a coaxial probe and shown in Fig. 3.8, is used to design a wideband-reconfigurable antenna. First, two similar E-shaped antennas (antennas 1 and 2) were optimized to operate at different frequencies, with their design parameters given in Table 3.5. Commercial software, HFSS 9.2, is used to simulate the characteristics of the E-shaped patches:

TABLE 3.4 Frequency-Reconfigurable Characteristics of the Reconfigurable Antenna Shown in Fig. 3.7

Case	Switch State	Operating Frequency Band (GHz) ($\lvert S_{11}\rvert < -10$ dB)	Bandwidth (%)
1	$1_6 2_6 3_8 4_8 5_8 6_4$	0.630–0.681	7.8
2	$1_6 2_6 3_8 4_7 5_7 6_3$	0.676–0.725	7.0
3	$1_6 2_6 3_7 4_7 5_7 6_3$	0.714–0.764	6.7
4	$1_6 2_6 3_7 4_6 5_6 6_3$	0.757–0.814	7.2
5	$1_6 2_6 3_6 4_6 5_6 6_4$	0.789–0.839	6.1
6	$1_5 2_5 3_6 4_5 5_6 6_3$	0.835–0.889	6.2
7	$1_5 2_5 3_5 4_5 5_5 6_4$	0.886–0.932	5.1
8	$1_4 2_4 3_4 4_5 5_5 6_2$	0.931–0.997	6.8
9	$1_4 2_4 3_4 4_5 5_4 6_4$	0.969–1.036	6.7
10	$1_3 2_3 3_4 4_3 5_3 6_1$	1.025–1.082	5.4
11	$1_3 2_3 3_3 4_3 5_3 6_1$	1.049–1.113	5.9
12	$1_2 2_2 3_3 4_2 5_2 6_1$	1.104–1.165	5.3
13	$1_2 2_2 3_2 4_2 5_2 6_1$	1.125–1.191	5.7
14	$1_1 2_1 3_1 4_1 5_1 6_1$	1.191–1.251	4.9
15	$1_0 2_0 3_0 4_0 5_0 6_0$	1.209–1.277	5.4

Source: [11].

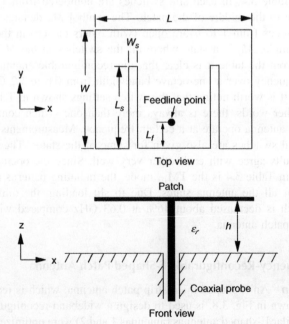

FIGURE 3.8 Configuration of an E-shaped patch antenna. (From [12], copyright © 2007 IET, with permission.)

TABLE 3.5 Design Parameters (Millimeters) of the Two E-Shaped Microstrip Patch Antennas, One for the Lower Band and Another for the Higher

Frequency	L	W	L_f	P_s	W_s	L_s	h	ε_r
Antenna 2	21.7	13.2	1.8	2.6	0.9	10.8	3	1
Antenna 1	14.7	9.9	1.8	1.6	0.9	8.4	3	1

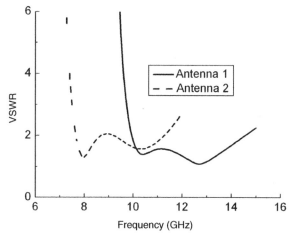

FIGURE 3.9 Simulated VSWRs for antennas 1 and 2. (From [12], copyright © 2007 IET, with permission.)

antennas 1 and 2. The simulated VSWRs of the two antennas are shown in Fig. 3.9. As can be seen from the figure, the simulated antenna bandwidths (VSWR ≤ 2) cover 9.9 to 14.6 GHz and 7.7 to 11.3 GHz, respectively. Special care is needed for the probe-feeding position and the slit dimensions so that the two antennas can be combined easily later. A foam (with thickness h_p and relative permittivity $\varepsilon_r = 1$) is selected to support the patch. The patch is fed by a 50-Ω coaxial probe of radius $D_p = 0.5$ mm at the small center patch with a ground plane size ($L_g \times W_g$) of 60×60 mm.

Then, with the use of 19 switches, a U-shaped patch is combined with three other rectangular patches to build a reconfigurable antenna that is switchable in two frequencies. The configuration is shown in Fig. 3.10. The design parameters are provided in Table 3.6. When all the shaded switches (with all the black switches in the OFF state) are turned on, the shape of the patch approximately equates to that for antenna 1, as can be seen from Fig. 3.11(a). On the other hand, the patch looks like antenna 2 when all the black switches are in the ON state [Fig. 3.11(b) with all the shaded switches turned OFF].

Ideal PIN diodes are used in all the simulations and measurements. In this case, the diode was simply replaced by a metallic pad (0.3×0.9 mm^2). This

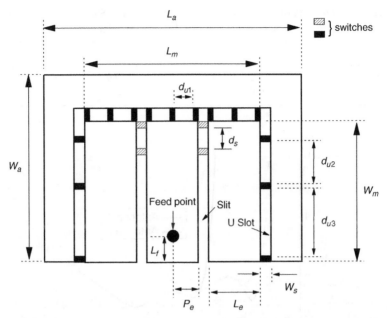

FIGURE 3.10 Configuration of a reconfigurable E-shaped patch antenna. (From [12], copyright © 2007 IET, with permission.)

TABLE 3.6 Optimized Design Parameters (Millimeters) of a Reconfigurable E-Shaped Patch

$L_a = 21.7$	$W_a = 31.2$	$L_m = 14.7$	$W_m = 9.9$
$h = 3.0$	$W_s = 0.9$	$P_e = 2.15$	$L_e = 4.3$
$L_f = 1.8$	$d_{u1} = 1.7$	$d_{u2} = 3.3$	$d_{u1} = 4.5$
$d_s = 1.8$	$\varepsilon_r = 1.0$	$D_p = 0.5$	$L_g = W_g = 60$

Source: [12].

is very useful for proof of concept. Of course, an actual PIN diode introduces additional losses and loading effects. Two E-shaped patch antennas were fabricated to resemble the two situations of the reconfigurable antenna; they are shown in Fig. 3.12. A copper plate with a thickness of 0.3 mm is selected to obtain good mechanical solidity. Some foam pieces with a relative permittivity of about 1 are used to support the patch. Figure 3.12(a) depicts a reconfigurable antenna where all the shaded switches are turned ON (all the black switches are OFF); whereas Fig. 3.12(b) is the antenna configuration with all the black switches turned ON (all the shaded switches are OFF).

Results and Discussion The simulated and measured VSWRs for two E-shaped patch antennas are shown in Fig. 3.13. In Fig. 3.13(a), the small patch (with all

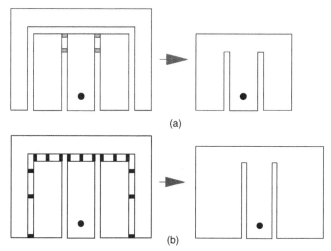

FIGURE 3.11   Shape of an E-shaped patch for two switch conditions: (a) all shaded switches ON (all black switches OFF); (b) all black switches ON (all shaded switches OFF).

FIGURE 3.12   Reconfigurable E-shaped patch antennas: (a) all shaded switches ON (all black switches OFF); (b) all black switches ON (all shaded switches OFF).

the black switches turned OFF and all the shaded switches turned ON) has three measured resonances at 9.3, 11.4, and 13.5 GHz, covering an impedance bandwidth (VSWR $\leq$ 2) of 38% (9.2 to 15 GHz). Figure 3.13(b) shows that the big patch (with all the black switches turned ON and all the shaded switches turned OFF) has two measured resonances at 7.8 and 9.8 GHz, with a bandwidth of 35% (7.5 to 10.7 GHz). For both cases, the discrepancy between simulation and measurement can be caused by mechanical tolerances. It was found by simulations that the width of the U-shaped slot and the positions of the diodes along the slot do not greatly affect antenna performance. Figures 3.14 and 3.15 show the simulated and measured radiation patterns of the second and third resonances (12.2 and 13.5 GHz) for the case when all the shaded switches are ON and all the black switches are

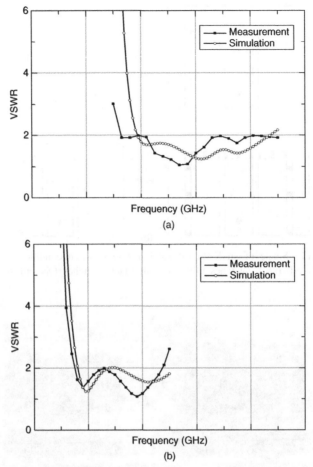

FIGURE 3.13 Simulated and measured VSWRs of a reconfigurable antenna with (a) all the shaded switches turned ON (all the black switches are OFF); (b) all the black switches turned ON (all the shaded switches are OFF). (From [12], copyright © 2007 IET, with permission.)

OFF. The first mode is similar to the second and is therefore omitted for brevity. Figures 3.16 and 3.17 give the simulated and measured radiation patterns of the first and second resonances (8 and 10.3 GHz) when all the shaded switches are OFF and all the black switches are ON. In general, good agreement between simulation and measurement is obtained for all cases.

3.4 PATTERN-RECONFIGURABLE ANTENNAS

A pattern-reconfigurable antenna is an EM radiator whose radiation patterns are changeable by switches. Usually, the operating frequency of such an antenna

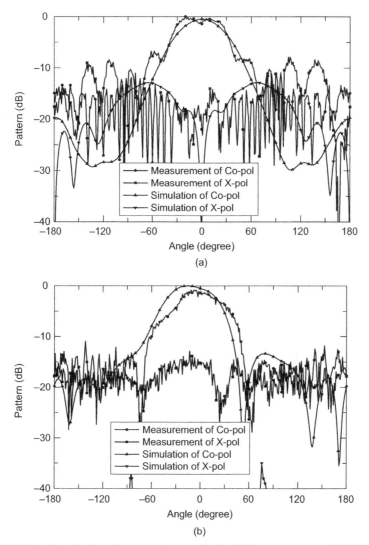

FIGURE 3.14   Simulated and measured normalized radiation patterns at 12.2 GHz (with all the shaded switches ON and all the black switches OFF): (a) xz-plane; (b) yz-plane. (From [12], copyright © 2007 IET, with permission.)

is kept constant. But in some cases it is designed to have different radiation patterns at different frequencies. Pattern reconfigurability enables wireless communication systems to avoid noisy environments, maneuver away from electronic jamming, and save energy by redirecting signals toward intended users only. Pattern-reconfigurable antennas are frequently used to increase the channel capacity and broaden the coverage area of a wireless system. Much work in past decades has been dedicated to studying pattern reconfigurable antennas.

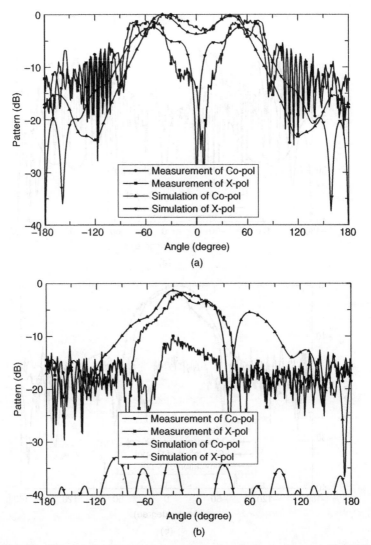

FIGURE 3.15 Simulated and measured normalized radiation patterns at 13.5 GHz (with all the shaded switches ON and all the black switches OFF): (a) xz-plane; (b) yz-plane. (From [12], copyright © 2007 IET, with permission.)

In 2003, Huff et al. [13] demonstrated a single-turn square-spiral microstrip antenna that can perform frequency and pattern reconfigurations simultaneously. Wu et al. [14] presented a pattern-reconfigurable microstrip patch antenna that has four switchable directional quasi-conical beams in different quadrants. To obtain higher antenna gain, Zhang et al. [15] and Yang et al. [16] introduced the Yagi–Uda pattern-reconfigurable microstrip parasitic arrays. For millimeter-wave applications, Xiao et al. [17,18] presented a novel co-planar-waveguide leaky-wave antenna with pattern reconfigurability.

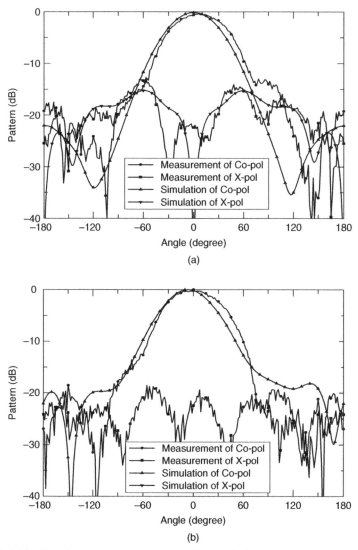

FIGURE 3.16 Simulated and measured normalized radiation patterns at 8 GHz (with all the shaded switches OFF and all the black switches ON): (a) xz-plane; (b) yz-plane. (From [12], copyright © 2007 IET, with permission.)

Pattern reconfiguration can also be accomplished by combining different antennas on a common structure [2]. In this case, different radiation patterns are usually excited simultaneously by altering the radiation aperture or by changing the feeding ports. Adding, removing, and shorting some parts of a radiation aperture are among the commonly used techniques [6,19]. For such antennas it is always very desirable to have a symmetrical radiation pattern. As a result, radiation elements

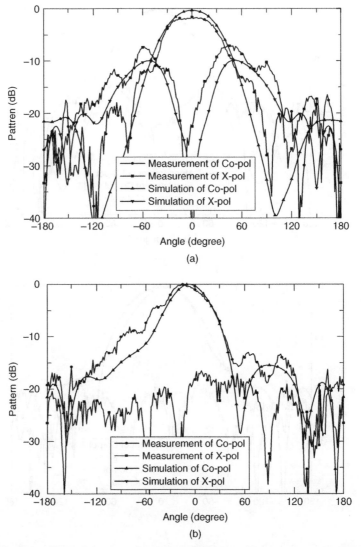

FIGURE 3.17 Simulated and measured normalized radiation patterns at 10.3 GHz (with all the shaded switches OFF and all the black switches ON): (a) xz-plane; (b) yz-plane. (From [12], copyright © 2007 IET, with permission.)

that have a symmetrical radiation aperture with a constant resonance frequency are preferable for designing pattern-reconfigurable antennas [15,19].

In this section, two design examples are examined. The first pattern-reconfigurable antenna is made on a fractal patch. In the second case, a traveling-wave antenna is made reconfigurable to achieve a narrow beamwidth and a wide scan range. Here, the beam directions can be scanned continuously by changing the period of the radiation element.

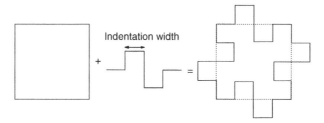

FIGURE 3.18 First-iteration square fractal shape.

3.4.1 Pattern-Reconfigurable Fractal Patch Antenna

Configuration The fractal patch antenna was introduced by Yeo and Mittra in 2001 [20]. A compact and pattern-reconfigurable square fractal loop was later proposed by Elkaruchouchi and El-Salam [21]. Figure 3.18 shows a first-iteration fractal shape that is built by combining a square and an indentation. Such a geometry is then employed in designing the pattern-reconfigurable planar microstrip fractal antenna [19] shown in Fig. 3.19. It has a top fractal patch and a metal ground plate. The top patch is 15×15 mm, the indentation width in the fractal generator is 2.5 mm, and the metal ground plate is 30×30 mm. The top square fractal patch area of the antenna is smaller than $0.5\lambda \times 0.5\lambda$ at 8.4 GHz. Four shorting pins are used to short the patch at four locations. One end of the pin is soldered to the ground and the other is connected by a switch to the edge of the fractal patch. Each shorting pin has dimensions of $1 \times 0.5 \times 4$ mm. The gap between the switch and the patch is 0.5 mm. The feedline consists of a vertical probe, a $\lambda/4$ microstrip impedance convertor, and a 50-Ω microstrip feedline. A vertical probe with a diameter of 1.27 mm is used to feed the patch through a gap of 0.15 mm, which is used to introduce capacitive coupling for better impedance matching. The vertical probe is connected to a $\lambda/4$ microstrip impedance convertor and a feedline, which are etched under the ground plate. A hole with diameter of 2 mm is milled on the ground plate to prevent the probe from touching the ground plate. In the experiment, a few foam boards ($\varepsilon_r \sim 1$) were used to support the patch and probe. The substrate under the ground plate has a thickness of 0.5 mm and a dielectric constant of $\varepsilon_r = 2.2$. Referring to Fig. 3.19, other parameters of the antenna are given as follows: $h = 4$ mm, $l_c = 8.5$ mm, $W_c = 3.5$ mm, and $W_f = 1.6$ mm.

By switching the four switches ON and OFF, the antenna can be set into four symmetrical settings. Each has only one switch turned ON. For ease of description, the switches are labeled numerically (1, 2, 3, and 4). Four groups of experiments (settings A, B, C, and D) are summarized in Table 3.7. For settings A and B, the xz-plane is designated as the E-plane as well as the scanning plane; the yz-plane is the H-plane. Similarly, for settings C and D, the yz-plane is the E-plane and also the scanning plane; the xz-plane is the H-plane.

Results and Discussion Ansoft HFSS was used to simulate the antennas. For proof of concept, all the switches are replaced by copper pads (with an area

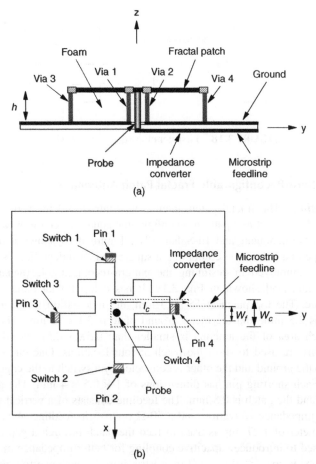

FIGURE 3.19 Configuration of a pattern-reconfigurable planar fractal antenna with switches: (a) front view; (b) top view.

of 1×0.5 mm) in both the simulations and measurements. Figure 3.20 shows the simulated and measured reflection coefficients in setting B. As can be seen from the figure, the measured resonance frequency of the fractal patch is about 8.4 GHz, with a measured impedance bandwidth of about 10%. The simulated result is very close to the measurement. Due to symmetry, all four antenna settings have almost the same reflection coefficient. In Fig. 3.21, the simulated and measured radiation patterns are given in the $E(xz)$- and $H(yz)$-planes. In the xz-plane, the antenna provides a half-power beamwidth ranging from 0 to 60° at an operating frequency of 8.4 GHz. Because of its symmetry with respect to the yz-plane, the radiation pattern at setting A has a half-power beamwidth covering 300 to 360° in the xz-plane. Similarly, settings C and D, respectively, provide half-power beamwidths of 300 to 360° and 0 to 60° in the yz-plane. As can be seen from Fig. 3.21, the sidelobes and cross-polarized fields of the four settings are less than

TABLE 3.7 **Switch Settings of a Pattern-Reconfigurable Fractal Antenna**

Setting	Switch 1	Switch 2	Switch 3	Switch 4
A	ON	OFF	OFF	OFF
B	OFF	ON	OFF	OFF
C	OFF	OFF	ON	OFF
D	OFF	OFF	OFF	ON

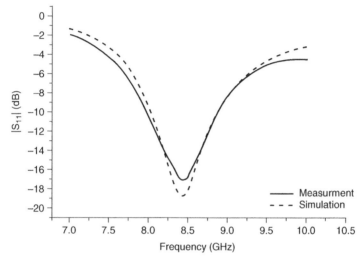

FIGURE 3.20 Simulated and measured reflection coefficients of a pattern-reconfigurable antenna in setting B. (From [19], copyright © 2007 IEEE, with permission.)

-6 and -10 dB, respectively. The simulated antenna gain is 4.29 dBi. If practical PIN diodes and RF MEMS devices are used for this application, the antenna gain can worsen at most by 1 dB at 8.4 GHz.

3.4.2 Pattern-Reconfigurable Leaky-Wave Antenna

Configuration Leaky-wave antennas are very popular because of a number of advantages, such as high directivity and compact size. In this section a pattern-reconfigurable co-planar waveguide (CPW) leaky-wave antenna is presented for millimeter-wave applications [17,18]. The configuration of a leaky-wave antenna is shown in Fig. 3.22. The longitudinal direction of the periodic structure is along the x-axis, and the design parameters are selected as follows: $W = 0.5$ mm, $S = 0.3$ mm, $h_1 = 0.794$ mm, $h_2 = 2$ mm, $L_s = 0.6$ mm, $W_s = 0.2$ mm, $\Delta P = 0.4$ mm, and $\varepsilon_r = 6.0$. The entire length of the antenna is $L = 50$ mm. Here, 125 pairs of perturbation slits (with a distance of ΔP) are etched on the inner edges of ground planes, aligned symmetrically with the origin O. One switch is installed at the open

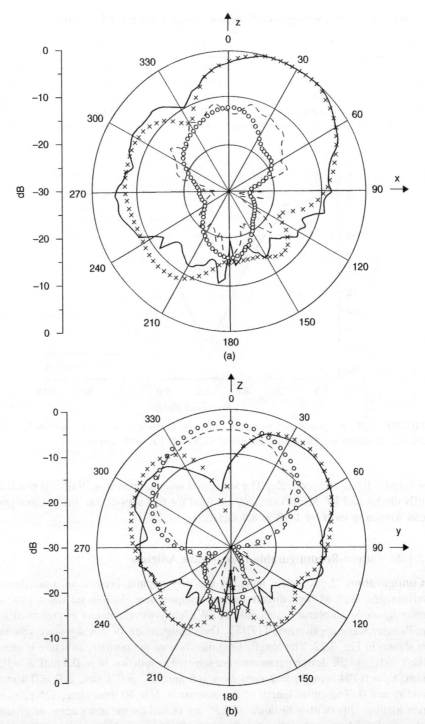

FIGURE 3.21 Simulated and measured radiation patterns of a pattern-reconfigurable antenna (setting B) at 8.4 GHz: (a) E-plane; (b) H-plane. (From [19], copyright © 2007 IEEE, with permission.)

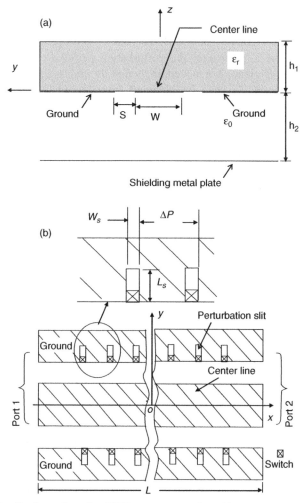

FIGURE 3.22 Configuration of a CPW pattern-reconfigurable antenna: (a) cross-sectional view; (b) CPW plane view. (From [18], copyright © 2005 IEEE, with permission.)

end of each perturbation slit. A perturbation period $P = p \times \Delta P$ (p is a positive integer) can be obtained by setting all p switches opened, with all others closed.

When the antenna is fed by port 1, the beam angle θ_m (the angle between the z-axis and the direction vector) of a spatial fast-wave component in the E-plane (xz-plane) is determined by [17]

$$\theta_m = \sin^{-1} \frac{\beta_{x0}}{k_0} - \frac{m\lambda_0}{P} \tag{3.1}$$

where, $\lambda_0, k_0, m, \beta_{x0}$, and P are the free-space wavelength, the free-space wave-number, the order of the spatial fast wave, the phase constant of the fundamental spatial harmonic inside the periodic structure, and the perturbation period, respectively. According to eq. (3.1), θ_m of the spatial fast wave can be adjusted by changing the perturbation period P. Therefore, by properly adjusting the states of switches in the slits to alter the perturbation period P, the antenna pattern can scan in the E-plane. For a certain period P and a single fixed frequency f_0, if β_{x0} is known, θ_m can be obtained from eq. (3.1). By combining the FDTD method with Floquet's theorem, the phase constant β_{x0} can easily be determined.

Results and Discussion   Four antenna settings with uniform periods of $P = 2.8$, 3.2, 3.6, and 4.0 mm, which correspond, respectively, to the open-switch settings of 7, 8, 9, and 10 switches, are selected to reconfigure the pattern. When the antenna is fed by port 1, the single main beam direction $\theta_m (m = 1)$ of the four settings can be predicted by eq. (3.1), and they are given in Table 3.8. From the table it can be seen that the single beam angle θ_m of the leaky-wave antenna can be changed within the range -90 to 0. However, a small-angle step scan cannot be achieved using only four uniform periods. A mixed-period structure is constructed to reduce the scanning angle step by cascading compound cells. The compound cell is formed by serially connecting the M cells with a period of P_1 and the N cells with a period of P_2.

The equivalent period of the mixed-period structure can be calculated from the equation

$$P_{\text{eff}} = \frac{MP_1 + NP_2}{M + N}, \tag{3.2}$$

where $P_1 - P_2 \leq \min(P_1, P_2)$. The phase constant of the mixed-period structure can be calculated by linear interpolation using β_{x0} of the two basic uniform periodic structures that compose the mixed-period structure. Once P_{eff} and the phase constant are obtained, the main beam angle θ_m of the mixed-period structure can be calculated using Eq. (3.1) $(m = 1)$. Table 3.9 shows the results of some mixed-period states of this reconfigurable leaky-wave antenna. Results obtained using the conventional FDTD method [18] are also given in the table.

TABLE 3.8   Electromagnetic Parameters of the Multi-reconfigurable Antenna Shown in Fig. 3.22

Period, P (mm)	β_{x0} by the New Method (rad/m)	θ_m by Eq. (3.1) (deg)	θ_m by the Conventional FDTD Method (deg)
2.8	1508	-90	-89
3.2	1473	-42	-42
3.6	1482	-21	-21
4.0	1506	-5.1	-6

Source: [17].

TABLE 3.9 Electromagnetic Parameters of Some Mixed-Period States of the Pattern-Reconfigurable Antenna Shown in Fig. 3.22

Basic Period (mm)	Compound Cell (M, N)	P_{eff} (mm)	β_{x0} (rad/m)	θ_m by Eq. (3.1)	θ_m by the Conventional FDTD Method (deg)
$P_1 = 2.8$ $P_2 = 3.2$	(2,1)	2.933	1496.3	−61.7	−65
	(1,1)	3.00	1490.5	−55.5	−58
	(1,2)	3.067	1484.7	−50.3	−51
$P_1 = 3.2$ $P_2 = 3.6$	(1,1)	3.40	1477.5	−30.4	−31
$P_1 = 3.6$ $P_2 = 4.0$	(1,1)	3.80	1494.0	−12.6	−16

Source: [17].

When the antenna is fed by port 1, but port 2 is loaded with a matched impedance, the single beam can scan from −90 to −1° by combining four uniform periodic states and five mixed periodic states. Due to the symmetry of the antenna structure, when the antenna is fed by port 2, the single beam can scan from 1 to 90° by using the nine states noted above.

3.5 MULTI-RECONFIGURABLE ANTENNAS

Configuration A multi-reconfigurable antenna can change more than one of its parameters such as frequency, pattern, polarization, and others. The ability to operate at different frequencies is highly desirable for cell phone and other wireless communication systems, which are required to work simultaneously at different standards. Pattern and polarization reconfigurations are frequently used to increase the channel capacity of a wireless system so that more users can share the same spectrum at the same time.

A Yagi–Uda antenna is able to provide both broad-side and end-fire radiation patterns. By introducing some slots and slits on its patch elements, a microstrip Yagi patch antenna was designed for frequency and pattern reconfigurations [22]. This antenna is shown in Fig. 3.23, with its detailed configuration depicted in Fig. 3.24. It can be made to radiate in different directions at different frequency bands. The antenna consists of a driven patch and four parasitic patch elements. For a Yagi–Uda antenna, a parasitic element can be made a director if it is smaller than the driven element; otherwise, it functions as a reflector. All of them are square metal patches printed on a grounded substrate, with dimensions of $50 \times 20 \times 1$ mm and a dielectric permittivity of 3. The driven element is 8×8 mm, approximately half the size of the guided wavelength at a resonance frequency of 9.5 GHz. According to the principle stated by Huang and Densmore [23], the dimension ratio of the director and driven elements should be made 0.8 to 0.95. All the parasitic elements have a uniform size of 7×7 mm, which is about 0.875 times that of the driven patch. The gaps between neighboring patches have the same

FIGURE 3.23 Multi-reconfigurable antenna. (From [22], copyright © 2008 John Wiley & Sons, Inc., with permission.)

distance: 0.8 mm. A 50-Ω coaxial probe is soldered along the array axis, with a distance of 1.7 mm away from the antenna center.

As can be seen from Fig. 3.24(a), two uniform slits are etched along the nonradiating edges of the driven patch, with a switch installed at the open end of each slit. As is well known, slits cause the resonance frequency to decrease. In this study, the slit has a length l of 2.3 mm and a width w_d of 0.4 mm, as shown in Fig. 3.24(b). The driven patch can be configured to have two different resonance frequencies by opening or closing the switches. When both switches are open, the antenna mode is termed the L mode. In this case, the resonance frequency (f_{dl}) of the driven patch decreases significantly. When both switches are closed, the antenna operates in the H mode. Now, the resonance frequency (f_{dh}) of the driven patch decreases slightly, with $f_{dh} > f_{dl}$. With reference to Fig. 3.24(c), a complex slot is etched on each parasitic patch to introduce extra inductance and capacitance. Also, three switches are installed along each slot to adjust the capacitance and inductance values. Three parasitic patch settings, B, S, and N, shown in Fig. 3.24(d), can be obtained. At setting B, all switches are open and the resonance frequency of the parasitic patch is much reduced. At setting S, only the middle switch is open; the others are closed. In this case, the resonance frequency of the parasitic decreases slightly. At setting N, all the switches are closed and the resonance frequency remains almost constant. The resonance frequencies of the parasitic element at these settings are called f_{pb}, f_{ps}, and f_{pn}, respectively, and we have $f_{pb} < f_{ps} < f_{pn}$. When it is in the H mode, the antenna settings are as follows:

- At setting B, the resonance frequency of the parasitic patch decreases, and it is lower than that of the driven patch: $f_{pb} < f_{dh}$. In this case, it acts as a reflector in the Yagi–Uda antenna.
- At setting S or N, the resonance frequency of the parasitic patch is still higher than that of the driven patch, and we have $f_{pn} > f_{ps} > f_{dh}$. The parasitic patch acts as a director.

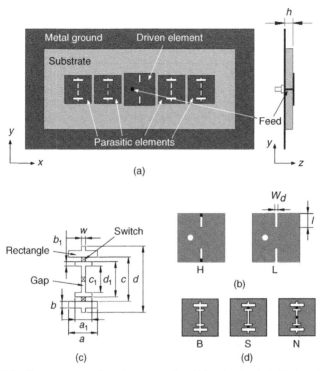

FIGURE 3.24   Frequency- and pattern-reconfigurable slot-loaded Yagi patch antenna: (a) antenna configuration; (b) driven patch in two frequency modes; (c) parameters of the slot: $w = 0.4$, $a = 2$, $b = 0.4$, $c = 3.6$, $d = 5.2$, $a_1 = 1.6$, $b_1 = 0.4$, $c_1 = 2$, and $d_1 = 2.8$ (units: mm); (d) three settings of the parasitic patch. The black squares are closed switches. (From [22], copyright © 2008 John Wiley & Sons, Inc., with permission.)

As no matching network is used when the antenna operates in the L mode, good impedance matching can only be obtained only if the resonance frequencies of the driven patch and the B parasitic element are close to each other. The roles of the parasitic elements are described as follows:

- At setting B, the resonance frequency of the parasitic element is higher but very close to that of the driven patch: $f_{pb} > \approx f_{dl}$. The parasitic patch acts as a director.
- At setting S or N, the resonance frequency of the parasitic element is much higher than that of the driven patch. Thus, it has a very weak directive effect on the radiation pattern.

Results and Discussion   By adjusting the settings of the parasitic patches, an antenna can be reconfigured in many different ways. When the parasitic patches on one side of the driver are directors while those on the other side are all reflectors,

TABLE 3.10 Simulated Data for the Five Configurations of the Multi-reconfigurable Antenna Shown in Fig. 3.24

Antenna Setting	Patch State	Bandwidth (GHz)	Beamwidth (deg)	Directivity
1	H-NNNN	9.02–9.46	−46 to 35	6.148
2	H-NBSS	9.13–9.64	6 to 57	7.887
3	H-SSBN	9.14–9.77	−61 to −7	7.703
4	L-NNBB	7.94–8.33	−4 to 53	7.783
5	L-BBNN	7.95–8.40	−53 to 6	7.706

Source: [22].

the main beam of the radiation pattern will tilt away from the broad side to the directors' side. When all the parasitic patches are directors or reflectors, a broad-side pattern can be obtained.

When the antenna operates in the H mode, by switching among the three configurations, the main beam of the antenna can cover a large continuous range in the upper half-space. When the antenna is in the L mode, by switching between two configurations, a large continuous beam coverage can also be achieved.

The simulated data for different settings of a pattern-reconfigurable antenna are compared in Table 3.10. For ease of description, the patches are labeled from left to right [refer to Fig.3.24(a)]. The summary of all settings, along with their simulated results, are also given in the table. The directivities and beamwidths of the H and L modes are obtained at 9.31 and 8.175 GHz, respectively. In this work, a metal pad 0.5×0.5 mm^2 is used to model the real switch for proof of concept.

The measured and simulated S parameters of the five configurations are shown in Fig. 3.25. Settings 1, 2, and 3, which belong to the H mode, cover an impedance passband ($|S_{11}| \leq -10$ dB) of 9.13 to 9.48 GHz; settings 4 and 5, which belong to the L mode, have a passband of 8.0 to 8.35 GHz. Simulated results show that the shared passband of settings 1, 2, and 3 is 9.14 to 9.46 GHz, and that for settings 4 and 5 is 7.95 to 8.33 GHz. The measured and simulated results agree well with each other.

The measured radiation patterns of the three H modes and two L modes at their central frequencies (9.31 and 8.175 GHz) are shown in Fig. 3.26. The main beams of settings 1, 2, and 3 cover a continuous angular range from −71 to +61° in the E-plane in the upper half-space. The two lower-frequency modes of settings 4 and 5 can make the radiation pattern cover a continuous radiation range from −56 to +56° in the E-plane.

3.6 CONCLUSIONS

In this chapter, several frequency-, pattern-, and multi-reconfigurable antennas have been discussed. Some recent work done by researchers at the Institute of Applied Physics, University of Electronic Science and Technology of China, Chengdu, has been presented. Here, both resonating and leaky-wave antennas are deployed for

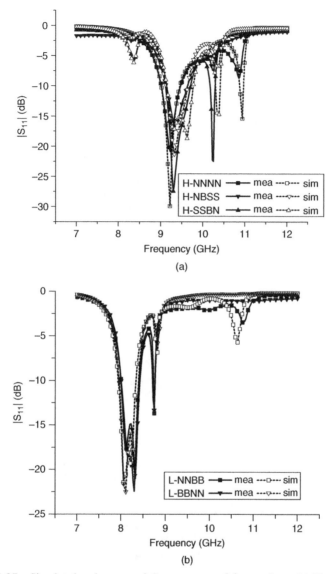

FIGURE 3.25 Simulated and measured S parameters of five settings: (a) H mode; (b) L mode. (From [22], copyright © 2008 John Wiley & Sons, Inc., with permission.)

designs of various reconfigurable antennas. Recent developments and issues have been described. Obviously, the development of reconfigurable antennas is always tied up with the improvement of semiconductor switches and diodes, which are used broadly to perform various antenna reconfigurations. With the availability of low-loss switches, the performance of reconfigurable antennas has been improved significantly over the past few years.

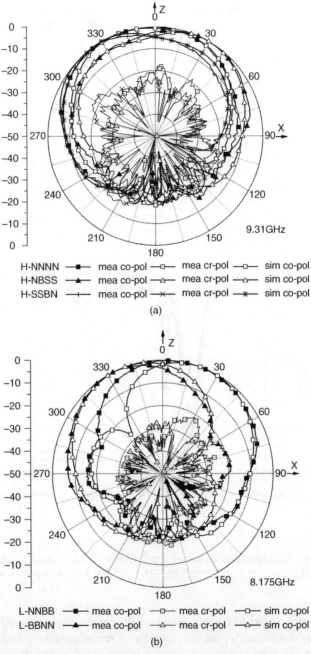

FIGURE 3.26 Measured and simulated radiation patterns in the *E*-plane of an antenna: (a) H mode; (b) L mode. (From [22], copyright © 2008 John Wiley & Sons, Inc., with permission.)

REFERENCES

[1] D. H. Schaubert, F. G. Farrar, S. T. Hayes, and A. Sindoris, "Frequency-agile, polarization diverse microstrip antennas and frequency scanned arrays," U.S. patent H01Q 001/38, 4367474, Jan. 4, 1983.

[2] W. S. Kang, J. A. Park, and Y. J. Yoon, "Simple reconfigurable antenna with radiation pattern", *Electron. Lett.*, vol. 44, no. 3, pp. 182–183, 2008.

[3] D. Peroulis, K. Sarabandi, and L. P. B. Katehi, "Design of reconfigurable slot antennas", *IEEE Trans. Antennas Propag.*, vol. 53, pp. 645–654, Feb. 2005.

[4] F. Yang and Y. Rahmat-Samii, "Patch antenna with switchable slot (PASS): dual-frequency operation", *Microwave Opt. Tech. Lett.*, vol. 31, no. 3, pp. 165–168, Nov. 2001.

[5] G. M. Rebeiz, *RF MEMS Theory, Design, and Technology*. Hoboken, NJ: Wiley, 2003.

[6] C. W. Jung, M.-J. Lee, G. P. Li, and F. De Flaviis, "Reconfigurable scan-beam single-arm spiral antenna integrated with RF-MEMS switches", *IEEE Trans. Antennas Propag.*, vol. 54, pp. 455–463, Feb. 2006.

[7] A. C. K. Mak, C. R. Rowell, R. D. Murch, and C.-L. Mak, "Reconfigurable multiband antenna designs for wireless communication devices", *IEEE Trans. Antennas Propag.*, vol. 55, pp. 1919–1928, July 2007.

[8] C. J. Panagamuwa, A. Chauraya, and J. C. Vardaxoglou, "Frequency and beam reconfigurable antenna using photoconducting switches", *IEEE Trans. Antennas Propag.*, vol. 54, pp. 449–454, Feb. 2006.

[9] K. C. Gupta, J. Li, R. Ramadoss, C. Wang, Y. C. Lee, and V. M. Bright, "Design of frequency-reconfigurable rectangular slot ring antennas", *IEEE Antennas and Propagation Society International Symposium*, no. 1, p. 326, 2000.

[10] P. F. Wahid, M. A. Ali, and B. C. DeLoach, Jr., "A reconfigurable Yagi antenna for wireless communications", *Microwave Opt. Tech. Lett.*, vol. 38, no. 2, pp. 140–141, July 2003.

[11] S. Xiao, B. Z. Wang, and X. S. Yang, "A novel frequency reconfigurable antenna", *Microwave Opt. Tech. Lett.*, vol. 36, no. 4, pp. 295–297, Feb. 2003.

[12] B. Z. Wang, S. Xiao and J. Wang, "Reconfigurable patch-antenna design for wideband wireless communication systems", *IET Microwave Antennas Propag.*, vol. 1, no. 2, pp. 414–419, Apr. 2007.

[13] G. H. Huff, J. Feng, S. Zhang, and J. T. Bernhard, "A novel radiation pattern and frequency reconfigurable single turn square spiral microstrip antenna", *IEEE Microwave Guided Wave Lett.*, vol. 13, no. 2, pp. 57–59, Feb. 2003.

[14] W. Wu, B. Z. Wang, and S. Sun, "Pattern reconfigurable microstrip patch antenna", *J. Electromagn. Waves Appl.*, vol. 19, no. 1, pp. 107–113, Jan. 2005.

[15] S. Zhang, G. H. Huff, J. Feng, and J. T. Bernhard, "A pattern reconfigurable microstrip parasitic array", *IEEE Trans. Antennas Propag.*, vol. 52, pp. 2773–2776, Oct. 2004.

[16] X. S. Yang, B. Z. Wang, and Y. Zhang, "Pattern reconfigurable quasi-Yagi microstrip antenna using photonic band gap structure", *Microwave Opt. Tech. Lett.*, vol. 42, no. 4, pp. 296–297, Aug. 2004.

[17] S. Xiao, B. Z. Wang, X. S. Yang, and G. Wang, "Novel reconfigurable CPW leaky-wave antenna for millimeter wave application", *Int. J. Infrared Millimeter Waves*, vol. 23, pp. 1637–1648, Nov. 2002.

[18] S. Xiao, Z. Shao, M. Fujise, and B. Z. Wang, "Pattern reconfigurable leaky-wave antenna design by FDTD method and Floquet's theorem", *IEEE Trans. Antennas Propag.*, vol. 53, pp. 1845–1848, May 2005.

[19] W. Wu, B. Z. Wang, X. S. Yang, and Y. Zhang, "A pattern-reconfigurable planar fractal antenna and its characteristic-mode analysis", *IEEE Antennas Propag. Mag.*, vol. 49, no. 3, pp. 68–75, June 2007.

[20] J. Yeo and R. Mittra, "Modified Sierpinski gasket patch antenna for multiband applications", *2001 IEEE International Symposium on Antennas and Propagation Digest*, vol. 3, pp. 134–137, June 2001.

[21] H. M. Elkaruchouchi and M. N. A. El-Salam, "Square loop antenna miniaturization using fractal geometry", *2003 IEEE International Symposium on Antennas and Propagation Digest*, vol. 4, pp. 22–27, June 2003.

[22] X. S. Yang, B. Z. Wang, S. Xiao, and K. F. Man, "Frequency and pattern reconfigurable Yagi patch antenna", *Microwavave Opt. Tech. Lett.*, vol. 50, no. 3, pp. 716–719, Mar. 2008.

[23] J. Huang and A. C. Densmore, "Microstrip Yagi array antenna for mobile satellite vehicle application", *IEEE Trans. Antennas Propag.*, vol. 39, pp. 1024–1030, July 1991.

CHAPTER FOUR

Receiving Amplifying Antennas

4.1 INTRODUCTION

As early as the 1960s, dipoles were integrated with low-noise amplifiers (LNAs) to enhance the signal-to-noise ratio (SNR) at receivers [1–3]. The new integrated antenna was called an *antennafier* because it combined the functions of radiation and amplification. Such integration was used to reduce the number of matching and tuning elements for a lower RF loss. When integrated with a transmitter [3], the power amplifying antenna was also called an antennamitter. In 1997, Radisic et al. [4] used a patch antenna as the harmonic tuning load of a power amplifier. By terminating the second harmonic of the power amplifier, the power-added efficiency (PAE) was found to be better. This method was later extended for suppressing the higher harmonics of a push–pull power amplifier [5]. Again, significant enhancement (with PAE > 55%) in the output power was observed. In the past two decades, antennas have been combined with both transmitters and receivers for designs of multifunctional transceivers [6,7]. Also, a lot of research has been reported on the integration of antenna and transceiver for quasi-optical spatial power combining during the millimeter-wave technology boom [8–11].

Obviously, the main advantage of integrating an antenna with an amplifier is the significant reduction in the length of signal transmission path that leads to improvement of signal quality and reduction of circuit size. Although fewer circuit elements are used, the impedance matching for the amplifying antennas remains very important. Since power amplifying antennas have been covered extensively by many publications and books (e.g., [1–11]), in this chapter we focus on discussing the matching techniques and new applications of low-noise amplifying

Compact Multifunctional Antennas for Wireless Systems, First Edition. Eng Hock Lim, Kwok Wa Leung.
© 2012 John Wiley & Sons, Inc. Published 2012 by John Wiley & Sons, Inc.

antennas. Co-design and co-optimization processes of such active antennas are also explored.

4.2   DESIGN CRITERIA AND CONSIDERATIONS

Figure 4.1 shows the functional blocks of an LNA, including its input- and output-matching circuits, at a typical receiver. Good stability, high gain, low noise, low power consumption, and low cost are among the important criteria of an LNA. To have unconditional stability, in amplifier design, it is crucial to bias a transistor so that the conditions stated in (4.1) and (4.2) are met. Condition (4.2) implies (4.3) and (4.4).

$$K = \frac{1 - |S_{11}|^2 - |S_{22}|^2 + |\Delta|^2}{2|S_{12}||S_{21}|} > 1 \qquad (4.1)$$

$$|\Delta| = |S_{11}S_{22} - S_{12}S_{21}| < 1 \qquad (4.2)$$

$$B_1 = 1 + |S_{11}|^2 - |S_{22}|^2 - |\Delta|^2 > 0 \qquad (4.3)$$

$$B_2 = 1 + |S_{22}|^2 - |S_{11}|^2 - |\Delta|^2 > 0 \qquad (4.4)$$

For an LNA, it is always very desirable to have a low noise factor F, which is defined as the ratio of the SNR at the input to the SNR at the output. The noise factor in decibels ($10 \log_{10} F$) is also called the *noise figure* (NF). Obviously, it is directly proportional to the noise that a transistor adds to the output signal. For a typical receiving active amplifying antenna, as shown in Fig. 4.2, the input-matching circuit is usually not needed. This can significantly reduce the footprint and circuit size.

4.3   WEARABLE LOW-NOISE AMPLIFYING ANTENNA

Configuration   In recent years, many passive wearable antennas have been reported [12–16]. Research was also conducted to study the effects of the

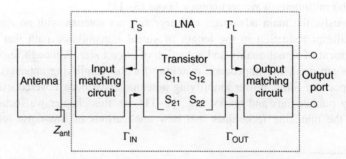

FIGURE 4.1   Block diagram of a low-noise amplifier at a typical receiver.

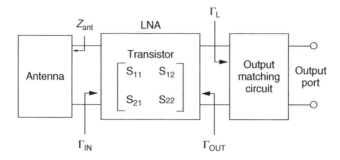

FIGURE 4.2 Block diagram of a receiving active amplifying antenna.

human body on such antennas [17,18]. In 2010, Declercq and Rogier [19] first proposed an active wearable antenna, made on garments. To demonstrate the design idea, an LNA is connected directly to a textile antenna without using an input-matching network. The impedance of the antenna is designed to match the input impedance of the LNA for optimal noise performance. In this case, the antenna impedance is designed to be coincident with the optimal impedance (of the transistor) for achieving a minimum noise figure. The schematic of the amplifier is shown in Fig. 4.3. It is designed using the biasing scheme recommended by Avago Technologies [20]. With the use of the power dividing resistors $R_1 = 300\ \Omega$, $R_2 = 1.2\ k\Omega$, and $R_3 = 10\ \Omega$, an Avago ATF-54143 + e-PHEMT transistor is biased in the common-source configuration with $V_{ds} = 3$ V, $I_d = 60$ mA, and $V_{gs} = 0.56$ V. A high-value resistor $R_4 = 10\ k\Omega$ has been used as the current limiter of the gate. Resistor $R_5 = 50\ \Omega$ functions as a resistive termination so that the low-frequency stability can be enhanced. The inductor $L_1 = 2.7$ nH and the $\lambda/4$ high-impedance (100-Ω) microstrip are both RF chokes,

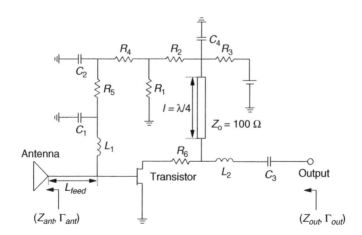

FIGURE 4.3 Schematic of an LNA [19].

while C_3 is a dc block. The components $C_1 = C_4 = 8.2$ pF and $C_2 = 10,000$ pF are bypass capacitors, where C_2 can further improve the low-frequency stability. An inductor L_2 is for output matching, while R_6 is for resistive loading. A flexible substrate made of a woven textile layer and a very thin flexible polyimid layer is used as a circuit board for accommodating the active devices and interconnects. The input port is connected directly to the receiving antenna through a via (diameter of 1 mm) to achieve low loss, low noise, and compact size.

Figure 4.4 shows the configuration of the ring antenna being used. It is built on a double-layered substrate and is designed to cover the entire 2.4 to 2.484 GHz ISM band. The top layer is made of a flexible polyurethane foam with a thickness of $h_1 = 3.56$ mm. As can be seen from the figure, the bottom layer consists of an aramid fabric (400 μm) and a polyimid textile (25 μm), together totaling to a substrate thickness of $h_2 = 425$ μm. The antenna and ground are made from a type of copper-plated nylon fabric called Flectron. All the materials are glued together by adhesive. The properties of the materials are given in Table 4.1. An active wearable antenna that has a total area of 11 cm^2 is shown in Fig. 4.5. As can be seen from the figure, the polyimid layer is kept as small as possible. Good bonding is formed between the aramid and polyimid layers.

The co-design and co-optimization processes of the antenna and amplifier require both field- and SPICE-based CAD softwares. In this case, the commercial software

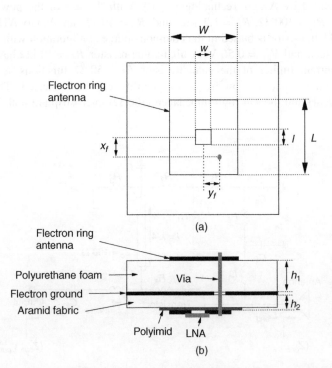

FIGURE 4.4   Configuration of an active wearable antenna: (a) top view; (b) front view.

TABLE 4.1 Properties of the Materials for an Active Wearable Antenna

	ε_r	tan δ	R_s at 2.45 GHz (Ω/sq)	σ (S/m)
Polyurethane foam	1.28	0.016		
Aramid fabric and polyimid textile	1.84	0.015		
Flectron	—	—	0.45	4.8×10^4

Source: [19].

Agilent ADS-Momentum and CST Microwave Studio are used jointly in designing the LNA and antenna, respectively. Material properties given in Table 4.1 have been used for all the simulations. To begin with, the LNA design is discussed. The transistor and all the lumped elements are simulated using Agilent's ADS. Momentum, a full-wave software affiliated with the ADS, can be employed for simulating the passive interconnects placed on the polyimid so that the parasitic effect can be accounted for. Then the antenna can be simulated using the field-based simulation software CST Microwave Studio, and the components can be combined using the following three-step co-design and co-optimization strategy [19]:

1. A transistor is first biased in its stable region to provide a low-noise figure and good linearity. It was found from the ADS-Momentum simulations that the LNA requires the antenna impedance to be $Z_{opt} = 35.015 + j17.644$ Ω for an achievable noise figure of $F_{min} = 0.715$ dB and a noise-matched power gain of 12.939 dB.

2. The CST Microwave Studio is used to search for a feeding position (x_f, y_f) on the ring antenna to yield an input impedance of $Z_{ant} = Z_{opt}$. The use of optimizer software eases the searching process. First, the 50-Ω feeding point of the antenna is found and the cost function is set to be $|\Gamma_{ant} - \Gamma_{opt}|$ in the optimizer. Then the optimizer is activated to automatically locate the feeding position that can meet this cost function.

3. Finally, the S parameters of the LNA are linked to the input impedance of the antenna using a dynamic link mechanism between the two softwares. With this link, the antenna parameters (L, W, l_{gap}, w_{gap}, x_f, and y_f) and LNA parameters (L_1, L_2, and L_{feed}) can be co-optimized simultaneously to meet the following specifications in the entire ISM band (2.4 to 2.48 GHz):

    (a) Noise figure that equals the minimum achievable noise figure of the LNA

    (b) Sufficient available gain G_A

    (c) Flat gain response $G_{A,max} - G_{A,min} < -10$ dB

    (d) Output matching $|\Gamma_{out}| < -10$ dB

As many solutions may meet the requirements, the simulated annealing was used as an optimization algorithm to find the global maximum. The co-optimized antenna and LNA parameters are shown in Tables 4.2 and 4.3. Also given in the

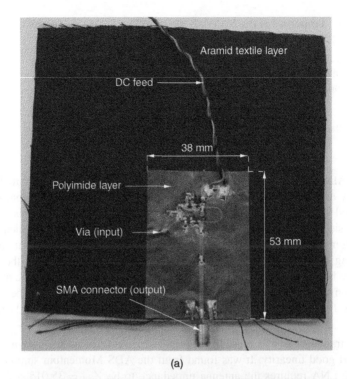

(a)

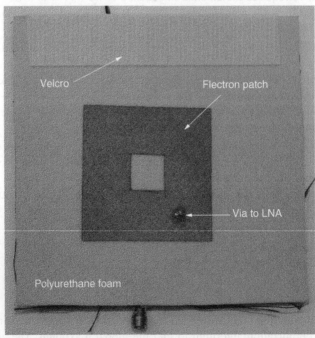

(b)

FIGURE 4.5 Active wearable antenna: (a) bottom view; (b) top view. (Courtesy of F. Declercq, Ghent University. From [19], copyright © 2010 IEEE, with permission.)

TABLE 4.2   Design Parameters of an Isolated and
Co-optimized Antenna

	Isolated Antenna	Co-optimized Antenna
L_{feed}	3 mm	4.73 mm
L_1	7.7 nH	10 nH
L_2	2.6 nH	1.2 nH

Source: [19].

TABLE 4.3   Design Parameters (Millimeters) of an
Isolated and Co-optimized LNA

	Isolated LNA	Co-optimized LNA
L	50.5	49.5
W	46.5	48
x_f	15	15.5
y_f	11	12
l_{gap}	9	13
w_{gap}	8	12

Source: [19].

tables are the design parameters for the isolated antenna and LNA. In general, the two groups of parameters are quite close to each other.

Results and Discussion

LNA Performance   The amplifier is first designed on the fabric materials without including the ring antenna. To measure the LNA output, the antenna is removed and the amplifier output is soldered to the center conductor of an SMA connector punching through the fabric. An Agilent N5242A PNA-X vector network analyzer (VNA) is used to measure the S parameters. The source-corrected noise measurement technique, which is a function affiliated to the PNA-X VNA, is adopted for measuring the noise figures [21]. To account for the effects of the human body, the LNA is placed on a person's back for measurements. With reference to Fig. 4.6, the simulated and measured amplifier gains ($|S_{21}|$) and noise figures are compared, with good agreement observed. The measured amplifier gain is 11.18 dB at 2.45 GHz, slightly lower than that for simulation (11.79 dB). As can be seen from the figure, the measured noise figure $F = 1.25$ dB is slightly higher than the simulated noise figure (0.87 dB). This on-body gain measurement shows that the active circuitry is not much affected by the human body despite it being directed toward the body without shielding. This is because the amplifier does not radiate and has little interaction with the human body. Figure 4.7 shows the simulated and measured stability factors K and B_1. In general, good agreement is observed, with the LNA meeting all the conditions for unconditional stability ($K > 1$ and $B_1 > 0$).

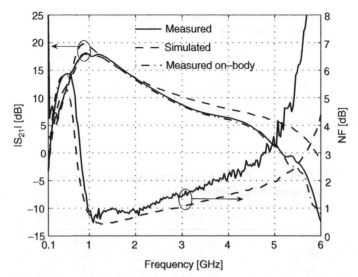

FIGURE 4.6 Simulated and measured amplifier gain and noise figure for the LNA in Fig. 4.3. (Courtesy of F. Declercq, Ghent University. From [19], copyright © 2010 IEEE, with permission.)

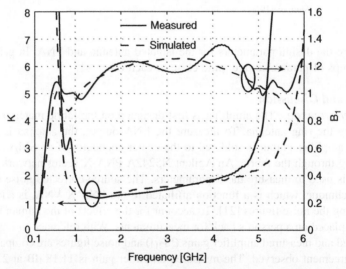

FIGURE 4.7 Simulated and measured stability factors K and B_1 for the LNA in Fig. 4.3. (Courtesy of F. Declercq, Ghent University. From [19], copyright © 2010 IEEE, with permission.)

Off-Body Active Antenna Performance

AMPLIFIER GAIN AND NOISE FIGURE MEASUREMENTS OF AN AMPLIFYING ANTENNA Now, the input port of the amplifier is connected to the optimized antenna. As the

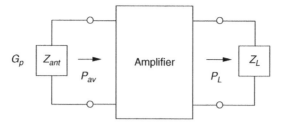

FIGURE 4.8 Configuration of a receiving amplifying antenna.

integrated antenna is now a one-port device, the conventional two-port method can no longer be used to measure the amplifier gain and noise figure. In 1994, An et al. [22] proposed a simple measurement method for receiving amplifying antennas. This method is adopted here. Figure 4.8 shows the configuration of a receiving active amplifying antenna. The transducer gain (G_T) is defined as

$$G_T = \frac{P_L}{P_{av}} \qquad (4.5)$$

where P_{av} is the available power received by the antenna (with input impedance Z_{ant}) and P_L is the power delivered to the load (with input impedance Z_L). The total available gain (G_A) of the active antenna can then be calculated:

$$G_{tot} = G_A = G_T G_p \qquad (4.6)$$

G_p, the antenna gain of the passive antenna, can be defined as

$$G_p = \frac{4\pi}{\lambda^2} A_e(\theta, \phi) \qquad (4.7)$$

where $A_e(\theta, \phi)$ is the effective aperture of the passive antenna.

The measurement is conducted inside an anechoic chamber, with the setup shown in Fig. 4.9. A transmitting standard gain horn is connected to port 1 of a VNA, and the receiving active amplifying antenna is connected to port 2. A similar passive antenna is also fabricated by removing the amplifier part. The distance (L) between the horn and the antenna under test must satisfy the far-field condition $(L > 10\lambda$ and $L > 2D^2/\lambda)$, where D is the maximum dimension of the antenna and λ is the operating wavelength. The transmission coefficients of the passive and active antennas are then measured, along with the reflection coefficient of the passive antenna. The Friis transmission equation states that S_{21a} transmitting from the horn to the active antenna can be expressed as

$$S_{21a} = 10 \log_{10} \frac{P_r}{P_t} = 10 \log_{10} \left(\frac{\lambda}{4\pi L}\right)^2 G_t G_{tot} = 10 \log_{10} \frac{G_t G_T A_e(\theta, \phi)}{4\pi L^2} \qquad (4.8)$$

where G_t is the antenna gain of of the standard gain horn.

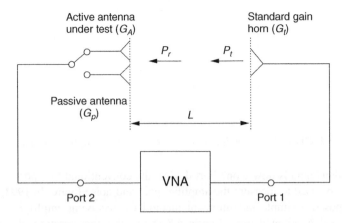

FIGURE 4.9 Measurement setup for amplifying antenna [22].

If the active antenna is replaced by the passive antenna,

$$S_{21p} = 10 \log_{10} \left(\frac{\lambda}{4\pi L} \right)^2 \left(1 - |S_{11p}|^2 \right) G_t G_p$$

$$= 10 \log_{10} \frac{\left(1 - |S_{11p}|^2 \right) G_t A_e(\theta, \phi)}{4\pi L^2} \qquad (4.9)$$

where S_{11p} is the reflection coefficient of the passive antenna. Subtracting (4.9) from (4.8), we obtain

$$S_{21a} - S_{21p} = 10 \log_{10} \frac{G_T}{1 - |S_{11p}|^2} \qquad (4.10)$$

Therefore,

$$G_T = S_{21a} - S_{21p} + 10 \log_{10}(1 - |S_{11p}|^2) \quad \text{(dB)} \qquad (4.11)$$

The transducer gain G_T can easily be calculated from (4.11) by knowing the reflection and transmission coefficients of the active and passive antennas.

Next, the noise figure measurement is discussed. With the knowledge of G_T, the noise factor of the receiving active amplifying antenna can be calculated using (4.12). The noise figure is defined as $\text{NF} = \log_{10} F$.

$$F = 1 + \frac{P_n}{G_T} - \frac{T_a}{290} \qquad (4.12)$$

where P_n is the absolute power of noise density that can be measured directly by a noise figure meter, and T_a is the ambient room temperature.

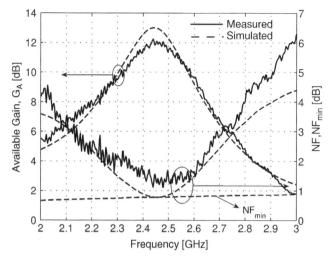

FIGURE 4.10   Simulated and measured NF and G_A of the active amplifying antenna in Fig. 4.5. (Courtesy of F. Declercq, Ghent University. From [19], copyright © 2010 IEEE, with permission.)

AVAILABLE GAIN AND NOISE FIGURE   In the measurement setup (Fig. 4.9), an Agilent N5242A PNA-X vector network analyzer (VNA) was used to measure the transmission (S_{21a}, S_{21p}) and reflection (S_{11p}) coefficients in an anechoic chamber. The absolute noise power density P_n was measured using the calibrated receiver of the PNA-X VNA. Measuring P_n at the output of the active antenna requires a long cable which has been included in the calibration. Inside the anechoic chamber, T_a is usually taken to be 300 K. The simulated and measured results are compared in Fig. 4.10, with good agreement. Using (4.6) and (4.11), the measured available gain G_A of the receiving active amplifying antenna is about 12 dB in the ISM frequency band (2.4 to 2.484 GHz), with a ripple amplitude of 0.5 dB. The simulated and measured noise figures are also given in Fig. 4.10 in the frequency range 2 to 3 GHz. An NF of about 1.3 dB is measured in the ISM band.

On-Body Active Antenna Performance   To further understand the effect of a human body on an antenna, an active antenna is placed inside a firefighter's jacket made of the same fabric materials. The antenna is placed at two positions (1 and 2) on the person's back [19]. For comparison, a similar passive antenna is measured at the same positions. Measurements were conducted in an anechoic chamber using the experimental setup described in Fig. 4.9. The free space and on-body total gains G_{tot} (or the total available gain G_A) of the active antenna were measured in the frequency range 2 to 3 GHz at both positions on the back, with the results shown in Fig. 4.11. Also depicted in the same figure are the antenna gains of the passive antenna at the two positions. The maximum free-space G_{tot} of the receiving amplifying antenna is about 17 dBi. As can be seen from the figure, human body has caused 1 to 3 dB of power loss to G_{tot}, depending on the position at which

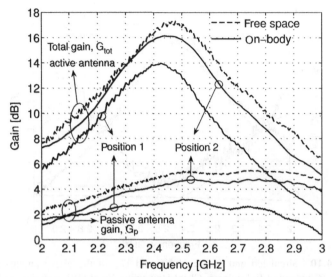

FIGURE 4.11 Measured gains of a receiving amplifying antenna at different body positions. (Courtesy of F. Declercq, Ghent University. From [19], copyright © 2010 IEEE, with permission.)

the antenna is placed on the body. It is also noted that the human body causes a noticeable degradation (about 1 to 3 dB) of the antenna gain. It is worth mentioning that position 1 has a higher loss because it is placed closer to the skin. Figure 4.12 shows the simulated and measured reflection coefficients in free space and on the body. As can be seen from the figure, the human body does not greatly affect the impedance matching. The discrepancy between the simulated and measured results can be caused by various experimental tolerances, which are quite significant in this case.

4.4 ACTIVE BROADBAND LOW-NOISE AMPLIFYING ANTENNA

Configuration In this section, the resistive equalization technique is used for co-designing and co-optimizing a broadband amplifying patch antenna. With the use of two equalizing resistors, the input impedance of a dual-band patch antenna can be configured to locate within the region where the field-effect transistor (FET) has the optimum noise figure. To demonstrate the design idea, a broadband receiving antenna is designed to cover the DCS-1800 and UMTS frequency bands.

Antenna Design The double-layered patch antenna reported by Rajo-Iglesias et al. [24] will be adopted for designing the following amplifying antenna. Figure 4.13 shows the antenna configuration, along with the design parameters: $W = 38$ mm, $h = 6$ mm, $d_1 = 10$ mm, $h = 6$ mm, and $\varepsilon_r = 3$. It has a ground size of 260×260 mm^2. Given in Fig. 4.14(a) is the measured input impedance of

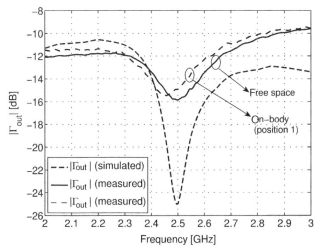

FIGURE 4.12   Simulated and measured reflection coefficients of the receiving amplifying antenna in Fig. 4.5 at various settings. (Courtesy of F. Declercq, Ghent University. From [19], copyright © 2010 IEEE, with permission.)

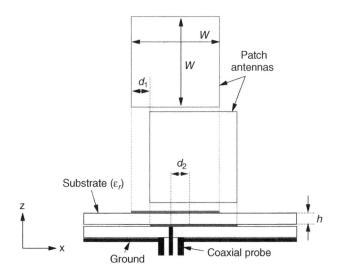

FIGURE 4.13   Double-layered wideband patch antenna [24].

the passive antenna with $d_2 = 4$ mm. As can be seen clearly from the figure, the antenna has two resonances, at 1.91 and 2.192 GHz. The antenna gain is about 7 ± 1.5 dB. Also shown on the same figure is the impedance region in which the FET has optimum noise (discussed in the next section). With reference to Fig. 4.14(a), unfortunately, the impedance locus of the antenna is far away from the optimum noise impedance region of the FET, which is very undesirable. It

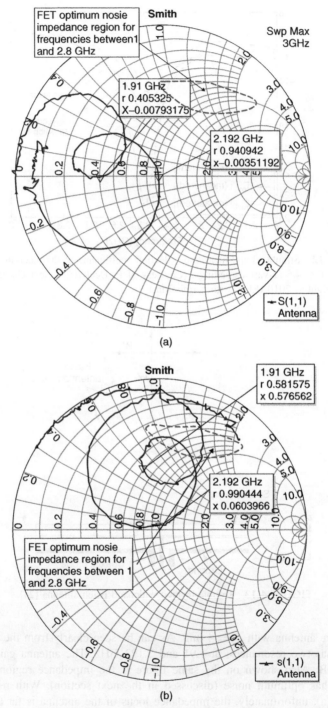

FIGURE 4.14   Input impedance of the double-layered antenna in Fig. 4.13 with (a) $d_2 = 4$ mm; (b) $d_2 = 9$ mm. (Courtesy of D. Segovia-Vargas, Carlos III University. From [23], copyright © 2008 IEEE, with permission.)

causes the amplifier to have a higher noise figure. By increasing the offset of the top–bottom patches to $d_2 = 9$ mm, shown in Fig. 4.14(b), a small portion of the impedance locus (around the resonances) can be moved into the optimum region. It will later be shown in the co-design process that the use of equalizing resistors can squeeze even more of the impedance locus of the antenna into the optimum noise impedance region.

Amplifier Design   The ATF34143 MESFET (Avagotech, San Jose, CA) is deployed for designing a low-noise amplifier. At a dc bias of $V_{ds} = 3$ V and $I_{ds} = 20$ mA [25], the transistor has a minimum noise figure of $F_{\min} = 0.17$ dB at 1.8 GHz, with the input port optimized to have $\Gamma_s = \Gamma_{\text{opt}} = 074\angle57°$. When applying the resistive equalization technique, the antenna is made to provide this impedance $\Gamma_{\text{ant}} = \Gamma_{\text{opt}}$. The small-signal model (inside the dashed square) of the FET is shown in Fig. 4.15, along with all the parasitics. As can be seen from the figure, g_m is the transconductance of the transistor. $R_g = R_{\text{gate}}$ and $R_d = R_{\text{drain}}$ are the external resistors that will be used for resistive equalization later.

Co-design of the Active Broadband Receiving Antenna   By connecting the antenna to the gate of the FET, a broadband amplifying antenna can be designed for the receiver, as shown in Fig. 4.16. The biasing voltages and current of the FET are $V_{ds} = 3$ V, $V_{gs} = -0.5$ V, and $I_{ds} = 20$ mA. The prototype (circuit side) is shown in Fig. 4.17. The antenna is built on the opposite side (not shown). It is obvious that no input-matching network is used for this active circuit. With reference to the figure, two resistors, R_g and R_d, have been added to the amplifier

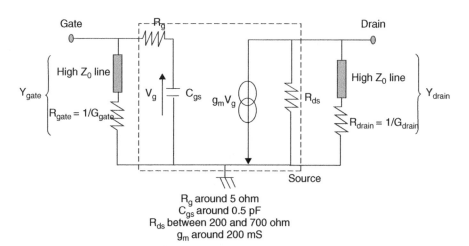

FIGURE 4.15 Equivalent circuit of an FET at low frequencies. (Courtesy of D. Segovia–Vargas, Carlos III University. From [23], copyright © 2008 IEEE, with permission.)

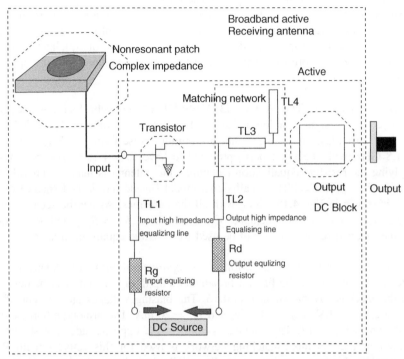

FIGURE 4.16 Configuration of a broadband amplifying antenna. (Courtesy of D. Segovia-Vargas, Carlos III University. From [23], copyright © 2008 IEEE, with permission.)

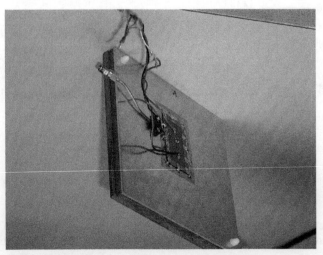

FIGURE 4.17 Broadband amplifying antenna. (Courtesy of D. Segovia-Vargas, Carlos III University. From [23], copyright © 2008 IEEE, with permission.)

to optimize the antenna efficiency and noise figure. Now, the resistive equalization technique is applied to meet the following design criteria:

1. Design the input impedance of the antenna to be a conjugate match $Z_{ant} = Z_{in}^*$ of that of the FET, or $\Gamma_{ant} = \Gamma_{in}^*$. At the same time, Z_{ant} (or Γ_{ant}) is made to locate within the optimum noise impedance region across the frequency $\Gamma_{ant} = \Gamma_{opt}$.
2. Make the transducer gain (G_T) as large as possible.
3. Provide good impedance matching at the amplifier output port.

As the input impedance of the antenna is complex and far from the resonance conditions, the generalized S parameters [26,27] can be used to analyze the network, shown in Fig. 4.18. Assuming that the input and output ports of the amplifier are ports 1 and 2, respectively, the matching condition at port 1 ($Z_{ant} = Z_{in}^*$) implies that

$$s_{11}^{GEN} \text{ and } Z_{IN} = f(Z_L, [S]) = Z_{ant}^* \tag{4.13}$$

where [S] denotes the S-parameter matrix of the FET.

By assuming that $Z_L = Z_0$ and $\Gamma_L = 0$ (Z_0 is the reference characteristic impedance), the generalized S parameters of the network in Fig. 4.18 can be written as

$$s_{11}^{GEN} = \frac{Z_{IN} - Z_{ant}}{Z_{IN} + Z_{ant}} = \frac{1}{1 - \Gamma_{ant} S_{11}} \frac{1 - \Gamma_{ant}}{1 - \Gamma_{ant}^*}(S_{11} - \Gamma_{ant}^*) \tag{4.14}$$

Conditions (4.13) and (4.14) imply that $s_{11}^* = \Gamma_{ant}$. Also, it is equal to the optimum noise impedance of the transistor, $\Gamma_{ant} = \Gamma_{opt}$. Considering the lossy small-signal model in Fig. 4.15, the inclusion of lossy shunt elements to the gate (Y_{gate}) and drain (Y_{drain}) of the FET leads to the new S_{11} expression.

$$S_{11} = \frac{[Y_0 - (Y_{11} + Y_{gate})][Y_0 + (Y_{22} + Y_{drain})] + Y_{12}Y_{21}}{[Y_0 + (Y_{11} + Y_{gate})][Y_0 + (Y_{22} + Y_{drain})] - Y_{12}Y_{21}} \tag{4.15}$$

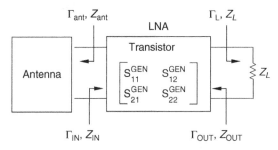

FIGURE 4.18 Generalized S parameters for a broadband amplifying antenna. (Courtesy of D. Segovia-Vargas, Carlos III University. From [23], copyright © 2008 IEEE, with permission.)

where Y_{ij} are the Y parameters of the FET. For the common case $G_{\text{drain}} \ll |Y_{21}|^2 R_n$, the minimum noise figure is given by

$$F_{\text{opt}} = F_{\text{min}}^{\text{total}} \approx 1 + 2R_n \left[(G_{\text{corr}} + G_{\text{gate}}) + 2\sqrt{\frac{G_{\text{corr}} + G_{\text{gate}}}{R_n} + (G_{\text{corr}} + G_{\text{gate}})^2} \right]$$

(4.16)

where R_n is the resistance of the equivalent noise voltage generator and G_{corr} is the correlated admittance between the voltage and current noise generators. The total noise figure can then be expressed as

$$F^{\text{total}} = F_{\text{min}}^{\text{total}} + \frac{R_n}{G_S}(G_S - G_{\text{opt}})^2 + \frac{R_n}{G_S}(B_S - B_{\text{opt}})^2 \qquad (4.17)$$

where G_{opt} and B_{opt} are defined as

$$G_{\text{opt}} = \sqrt{\frac{G_n + G_{\text{gate}}}{R_n} + (G_{\text{corr}} + G_{\text{gate}})^2} \quad \text{and} \quad B_{\text{opt}} = -(B_{\text{corr}} + B_{\text{gate}}) \quad (4.18)$$

It can be seen from eq. (4.18) that the noise figure is affected by the gate conductance. The effect of gate conductance is studied in Fig. 4.19. As can be seen from the figure, a higher R_g/Z_0 value is good for a lower noise figure.

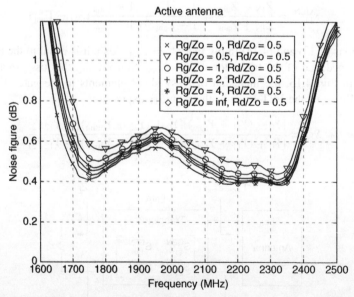

FIGURE 4.19 Effect of the gate conductance on F^{total} across the frequency. (Courtesy of D. Segovia-Vargas, Carlos III University. From [23], copyright © 2008 IEEE, with permission.)

Using the model in Fig. 4.18, the transducer gain (G_T) can be defined as

$$s_{12}^{\text{GEN}} = \frac{1}{1 - \Gamma_{\text{ant}} S_{11}} \frac{1 - \Gamma_{\text{ant}}}{|1 - \Gamma_{\text{ant}}|} S_{12} (1 - |\Gamma_{\text{ant}}|^2)^{\frac{1}{2}} \qquad (4.19)$$

$$G_T = s_{21}^{\text{GEN}} = \frac{S_{21}}{S_{12}} s_{12}^{\text{GEN}} \qquad (4.20)$$

where the transmission coefficient S_{21} is defined as

$$S_{21} = \frac{-2g_m Z_0}{[1 + G_{\text{gate}} Z_0][1 + (G_{da} + G_{\text{drain}}) Z_0]} \qquad (4.21)$$

Results and Discussion The AWR Microwave Office was used to optimize the models in Figs. 4.15 and 4.16, and it was found that the optimized drain and gate resistances are $R_d = R_{\text{drain}} = 91\ \Omega$ and $R_g = R_{\text{gate}} = 68\ \Omega$, respectively. With reference to Fig. 4.16, two high-impedance lines ($Z_0 = 90\ \Omega$, with an electrical length of $90°$) are used to compensate the effect of losses at higher frequencies. The new input impedance of the antenna (inclusive of R_d and R_g) is shown in Fig. 4.20. The addition of the two resistors causes the locus of the antenna impedance to become narrower. Therefore, a larger portion of the locus can then be squeezed into the region in which the FET has optimum noise. Also plotted on the figure is the locus of the optimum noise impedance of the FET. It can be seen from the figure that better impedance matching can now be obtained as the two loci are closer to each other [compared to Fig. 4.14(b)]. These two resistors are called *equalizing resistors*.

Now, the gain of the active antenna is characterized. For an active amplifying antenna, its effective transmission gain (G_{TX}) can be obtained by comparing its transmission coefficient to that of its passive counterpart. Of course, G_{TX} is a figure that combines the antenna and amplifier gains. A standard reference horn is used for measurement in an anechoic chamber. With reference to Fig. 4.21, the measured $G_{\text{TX}}(= \Delta S_{21})$ reads about 13 dB across the frequency passband. It is then compared with the simulated G_T of the isolated amplifier in Fig. 4.22. It is worth mentioning that the antenna impedance is used as the port impedance at the input. As can be seen from the figure, the simulated G_T agrees reasonably well with the measured G_{TX}.

The ratio of antenna gain to effective system temperature (G/T) ratio is a figure of merit that describes the performance of an AIA [28,29]. For an active antenna, its G and T cannot be measured separately. The G/T ratio is defined as

$$\frac{G}{T} = \frac{G_{\text{ACT}}/\Delta G}{T_a + T_0(L_e - 1) + T_0(F_{\text{opt}} - 1)L_e + T_0(L - 1)\frac{L_e}{\Delta G \cdot L} + \frac{T_{\text{rec}} \cdot L_e}{\Delta G}} \qquad (4.22)$$

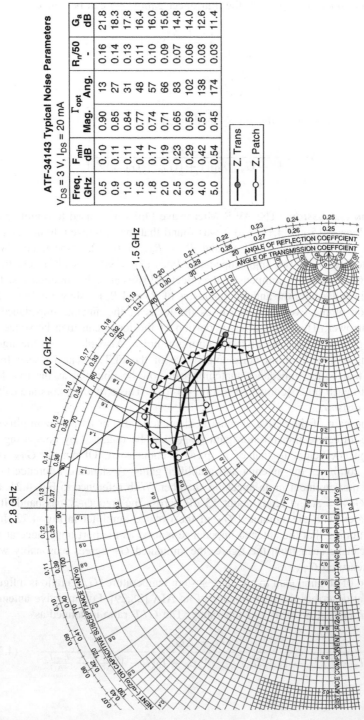

ATF-34143 Typical Noise Parameters

$V_{DS} = 3$ V, $I_{DS} = 20$ mA

Freq. GHz	F_{min} dB	Γ_{opt} Mag.	Γ_{opt} Ang.	$R_n/50$ -	G_a dB
0.5	0.10	0.90	13	0.16	21.8
0.9	0.11	0.85	27	0.14	18.3
1.0	0.11	0.84	31	0.13	17.8
1.5	0.14	0.77	48	0.11	16.4
1.8	0.17	0.74	57	0.10	16.0
2.0	0.19	0.71	66	0.09	15.6
2.5	0.23	0.65	83	0.07	14.8
3.0	0.29	0.59	102	0.06	14.0
4.0	0.42	0.51	138	0.03	12.6
5.0	0.54	0.45	174	0.03	11.4

Z. Trans
Z. Patch

FIGURE 4.20 New input impedance of a double-layered patch antenna (with inclusion of R_d and R_g). Also given is the locus of the optimum noise impedance of the FET. *Z. Trans* refers to the optimum noise impedance of the transistor and *Z. Patch* refers to the new input impedance of the antenna. (Courtesy of D. Segovia-Vargas, Carlos III University. From [23], copyright © 2008 IEEE, with permission.)

136

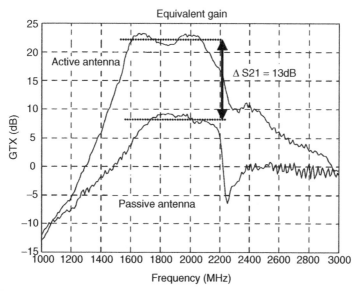

FIGURE 4.21 Measured effective transmission gain (G_{TX}) of active and passive antennas. (Courtesy of D. Segovia-Vargas, Carlos III University. From [23], copyright © 2008 IEEE, with permission.)

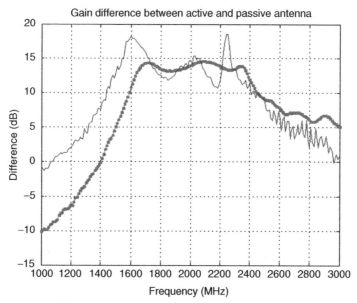

FIGURE 4.22 Comparison of the simulated transducer gain (G_T) and the measured effective transmission gain (G_{TX}). (Courtesy of D. Segovia-Vargas, Carlos III University. From [23], copyright © 2008 IEEE, with permission.)

with the following parameters:

G_{ACT} antenna gain of the active antenna
ΔG total gain, including antenna gain, amplifier gain, and losses
L_e loss factor (>1)
L losses
T_0 physical temperature of the antenna
T_a temperature of the active antenna
T_{re} receiver equivalent noise temperature

Once G/T is known, the total noise figure of the active antenna can be calculated:

$$\text{NF} = F^{\text{total}} = F_{\text{opt}} \cdot L_e + \frac{(L-1)L_e}{L \Delta G} \tag{4.23}$$

Figure 4.23(a) shows the measurement setup for the active antenna in an anechoic chamber. With reference to the figure, a section of 50-Ω transmission line (with a loss of 2 dB) is used to connect the active antenna to the receiver (with a loss of 8 dB). The passive antenna can be measured by the setup in Fig. 4.23(b). In this case, the antenna is connected to an external LNA through a section of transmission line (with a loss of 2 dB). Figure 4.24 shows the measured G/T of the active antenna for different (V_{ds}, V_{gs}) biasing conditions. Several combinations (1.5V, 0.5V), (2.5V, 0.5V), (3.0V, 0.6V), and (3.5V, 0.6V) were tested. Also shown on the figure is the measured G/T of the passive antenna. When biased at $V_{ds} = 2.5$ V and $V_{gs} = 0.5$ V, the ratio G/T is about -16 dB/K in the frequency range 1.5 to 2.05 GHz (bandwidth of $\sim$30%), as can be seen from Fig. 4.24. It is 4 to 6 dB better than that for the passive antenna.

By letting the antenna temperature $T_0 = T_a = 290$ K and knowing the G/T ratio, the noise figure NF can be calculated using (4.23). Figure 4.25 shows the NF calculated from the simulated and measured G/T values. With reference to the figure, the NF measured at 1.8 GHz is about 0.5 dB, being higher than that given in the datasheet (0.17 dB).

Figure 4.26 shows the simulated and measured reflection coefficients of the broadband amplifying antenna. Since port 1 of the amplifier is now connected to the antenna, the S_{22} notation is used to denote the reflection coefficient at the new input port of the active antenna. Three resonances are observed in both the simulation and measurement, with reasonable agreement. Also shown in the figure is the measured reflection coefficient of the passive antenna. It is higher because the antenna works in the nonresonant modes. Figure 4.27 shows the radiation patterns measured for the active and passive antennas at 1.8 and 2.2 GHz, both in the E-plane. They have almost the same field patterns, implying that the amplifier does not greatly disturb the far-field characteristics of the passive antenna.

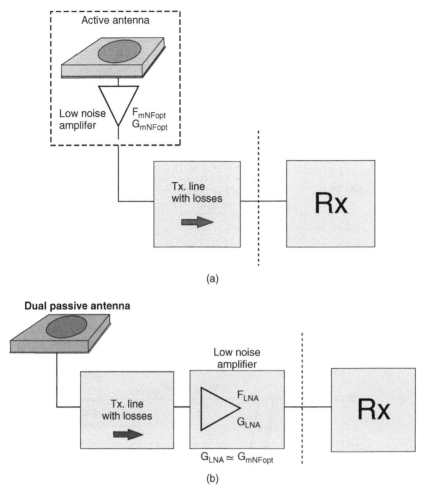

FIGURE 4.23 Measurement setups for (a) an active antenna; (b) a passive antenna. (Courtesy of D. Segovia-Vargas, Carlos III University. From [23], copyright © 2008 IEEE, with permission.)

4.5 CONCLUSIONS

In this chapter, the co-design and co-optimization processes of receiving amplifying antennas have been explored. In the first part, the amplifying antenna is made on textile and is wearable. The interaction between an antenna, an LNA, and the human body has been studied. It has been found in the second part that wide impedance matching is achievable by deploying two equalization resistors to the gate and drain of a FET. For both parts, the antenna performance is not greatly affected by the amplifier.

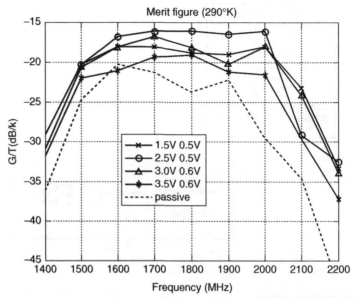

FIGURE 4.24 Measured G/T of the active antenna in Fig. 4.16 with different biasing conditions (V_{ds} and V_{gs}). Also shown is the measured G/T of a passive antenna. (Courtesy of D. Segovia-Vargas, Carlos III University. From [23], copyright © 2008 IEEE, with permission.)

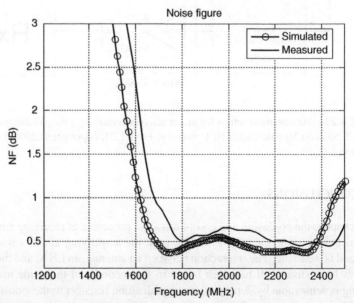

FIGURE 4.25 Comparison of noise figures (NFs) calculated from simulated and measured G/T. (Courtesy of D. Segovia-Vargas, Carlos III University. From [23], copyright © 2008 IEEE, with permission.)

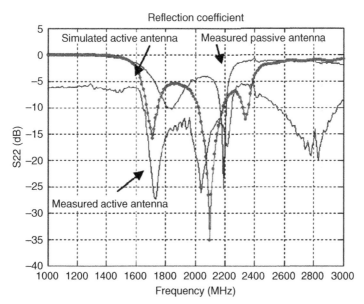

FIGURE 4.26 Simulated and measured reflection coefficients (S_{22}) of the active antenna in Fig. 4.16. Also shown is the S_{22} measured for a passive antenna. (Courtesy of D. Segovia-Vargas, Carlos III University. From [23], copyright © 2008 IEEE, with permission.)

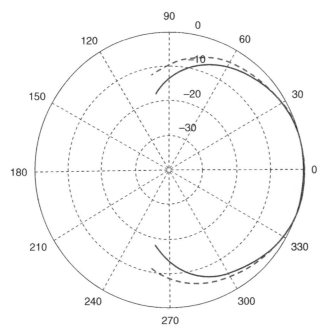

E plane (1.8 GHz)-Active and passive patch radiation pattern

(a)

FIGURE 4.27 Co-polarized fields measured for active (solid line) and passive (dashed line) antennas in the E-plane. (Courtesy of D. Segovia-Vargas, Carlos III University. From [23], copyright © 2008 IEEE, with permission.)

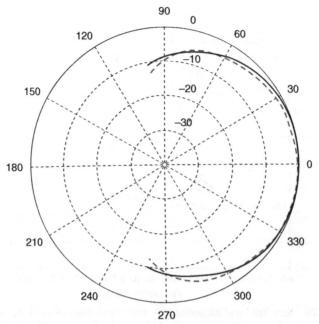

E plane (2.2 GHz)-Active and passive patch radiation pattern

(b)

FIGURE 4.27 (*Continued*)

REFERENCES

[1] A. D. Frost, "Parametric amplifier antenna," *Proc. IRE*, vol. 48, pp. 1163–1164, June 1960.

[2] A. D. Frost, "Parametric amplifier antenna," *IEEE Trans. Antennas Propag.*, vol. 12, no. 2, pp. 234–235, Mar. 1964.

[3] J. R. Copeland, W. J. Robertson, and R. G. Verstraete, "Antennafier arrays," *IEEE Trans. Antennas Propag.*, vol. 12, no. 2, pp. 227–233, Mar. 1964.

[4] V. Radisic, S. T. Chew, Y. X. Qian, and T. Itoh, "High-efficiency power amplifier integrated with antenna," *IEEE Microwave Guided Wave Lett.*, vol. 7, pp. 39–41, Feb. 1997.

[5] W. R. Deal, V. Radisic, Y. X. Qian, and T. Itoh, "Novel push–pull integrated antenna transmitter front-end," *IEEE Microwave Guided Wave Lett.*, vol. 8, no. 11, pp. 405–407, Nov. 1998.

[6] M. J. Cryan, P. S. Hall, S. H. Tsang, and J. Sha, "Integrated active antenna with full duplex operation," *IEEE Trans. Microwave Theory Tech.*, vol. 45, pp. 1742–1748, Oct. 1997.

[7] Y. Chung, S. S. Jeon, S. Kim, D. Ahn, J. I. Choi, and T. Itoh, "Multifunctional microstrip transmission lines integrated with defected ground structure for RF front-end application," *IEEE Trans. Microwave Theory Tech.*, vol. 52, pp. 1425–1432, May 2004.

[8] J. A. Navarro and K. Chang, *Integrated Active Antennas and Spatial Power Combining*. New York: Wiley, 1996.

[9] A. Mortazawi, T. Itoh, and J. Harvey, *Active Antennas and Quasi-optical Arrays*. Piscataway, NJ: IEEE Press, 1998.

[10] R. A. York and Z. Popovic, Eds., *Active and Quasi-optical Arrays for Solid-State Power Combining*. New York: Wiley, 1997.

[11] M. P. Delisio and R. A. York, "Quasi-optical and spatial power combining," *IEEE Trans. Microwave Theory Tech.*, vol. 50, pp. 929–936, Mar. 2002.

[12] P. O. Salonen, Y. Rahmat-Samii, H. Hurme, and M. Kivikoski, "Dualband wearable textile antenna," *Proceedings of the IEEE Antennas and Propagation International Symposiun*, vol. 1, pp. 463–467, 2004.

[13] A. Tronquo, H. Rogier, C. Hertleer, and L. Van Langenhove, "Robust planar textile antenna for wireless body LANs operating in 2.45GHz ISM band," *Electron. Lett.*, vol. 42, no. 3, pp. 142–143, Feb. 2006.

[14] I. Locher, M. Klemm, T. Kirstein, and G. Tröster, "Design and characterization of purely textile patch antennas," *IEEE Trans. Adv. Packag.*, vol. 29, pp. 777–788, Nov. 2006.

[15] T. F. Kennedy, P. W. Fink, A. W. Chu, and G. F. Studor, "Potential space applications for body-centric wireless and e-textile antennas," *Proceedings of the IET Seminar on Antennas, Propagstion and Body-Centric Wireless Communications*, pp. 77–83, Apr. 2007.

[16] C. Hertleer, H. Rogier, L. Vallozzi, and L. Van Langenhove, "A textile antenna for off-body communication integrated into protective clothing for fire-fighters," *IEEE Trans. Antennas Propag.*, vol. 57, pp. 919–925, Apr. 2009.

[17] P. S. Hall, Y. Hao, Y. I. Nechayev, A. Alomainy, C. C. Constantinou, C. Parini, M. R. Kamarudin, T. Z. Salim, D. T. M. Hee, R. Dubrovka, A. S. Owadally, W. Song, A. Serra, P. Nepa, M. Gallo, and M. Bozzetti, "Antennas and propagation for on-body communication systems," *IEEE Antennas Propag. Mag.*, vol. 49, no. 3, pp. 41–58, June 2007.

[18] G. A. Conway and W. G. Scanlon, "Antennas for over-body-surface communication at 2.45GHz," *IEEE Trans. Antennas Propag.*, vol. 57, pp. 844–855, Apr. 2009.

[19] F. Declercq and H. Rogier, "Active integrated wearable textile antenna with optimized noise characteristics," *IEEE Trans. Antennas Propag.*, vol. 58, pp. 3050–3054, Sept. 2010.

[20] ATF-54143: High Intercept Low Noise Amplifier for the 1850–1910MHz PCS Band Using the Enhancement Mode PHEMT. Application Note 1222, Avago Technologies, 2006.

[21] K. Wong, Advancements in Noise Measurement. http://www.ewh.ieee.org/r6/scv/ims/archives/May2008Wong.pdf

[22] H. An, B. K. J. C. Nauwelaers, A. R. Van de Capelle, and R. G. Bosisio, "A novel measurement technique for amplifier-type active antennas," *IEEE MTT-S International Symposium Digest*, vol. 3, pp. 1473–1476, 1994.

[23] D. Segovia-Vargas, D. Castro-Galán, L. E. García-Mu noz, and V. González-Posadas, "Broadband active receiving patch with resistive equalization," *IEEE Trans. Microwave Theory Tech.*, vol. 56, pp. 56–64, Jan. 2008.

[24] E. Rajo-Iglesias, D. Segovia-Vargas, J. L. Vázquez-Roy, V. González-Posadas, and C. Martín-Pascual, "Bandwidth enhancement in noncentered stacked patches," *Microwave Opt. Tech. Lett.*, vol. 31, no. 1, pp. 32–34, Oct. 2001.

[25] ATF-34143: Low Noise Pseudomorphic HEMT in a Surface Mount Plastic Package. Avago Technologies, 2009.

[26] R. Collin, *Foundations for Microwave Engineering*, 2nd ed. New York: McGraw-Hill, 1992.

[27] K. Kurokawa, "Power wave and the scattering matrix," *IEEE Trans. Microwave Theory Tech.*, vol. 13, pp. 194–202, Mar. 1965.

[28] J. J. Lee, "G/T and noise figure of active array antennas," *IEEE Trans. Antennas Propag.*, vol. 41, pp. 241–244, Feb. 1993.

[29] U. R. Kraft, "Gain and G/T of multielement receive antennas with active beamforming networks," *IEEE Trans. Antennas Propag.*, vol. 48, pp. 1818–1829, Dec. 2000.

CHAPTER FIVE

Oscillating Antennas

5.1 INTRODUCTION

A microwave oscillator is a device that converts direct-current (dc) energy into alternating-current (ac) energy [1]. In modern wireless systems, microwave oscillators are commonly used to generate carrier signals for radio transmitters. At RF receivers, oscillators are usually associated with mixers and phase-locked loops for extracting message signals. For decades, active devices such as Gunn diodes, impact ionization avalanche transit time (IMPATT) devices, and resonant tunneling diodes (RTDs) have been used broadly to perform this important dc-to-ac conversion. Bipolar-junction transistors (BJT), metal–semiconductor field-effect transistors (MESFET), metal-oxide-semiconductor field-effect transistors (MOS-FET), and high-electron-mobility transistors (HEMT) [2] are among the modern transistors in common use in the design of various oscillator circuits.

In past decades, oscillators were integrated with antennas mainly for spatial power combining [1,3]. This is because a microstrip is extremely lossy in the millimeter-wave ranges, while, on the other hand, a spatial power combining has relatively lower loss. An oscillating antenna, also called an *antenna oscillator*, is a multifunctional circuit that can be designed easily by employing an antenna simultaneously as the resonator, the feedback element, and as the load of an oscillator. In this chapter, design examples are given for all the aforementioned cases.

5.2 DESIGN METHODS FOR MICROWAVE OSCILLATORS

There are many ways to design a microwave oscillator. In this section, attention is given to S parameters and the network model since most antenna oscillators can be analyzed conveniently using these two methods.

Compact Multifunctional Antennas for Wireless Systems, First Edition. Eng Hock Lim, Kwok Wa Leung.
© 2012 John Wiley & Sons, Inc. Published 2012 by John Wiley & Sons, Inc.

5.2.1 Design Using S Parameters

Design using S parameters has already been well covered by many research papers and textbooks [2,4–6]. We discuss only briefly the design procedure of an oscillator based on the small-signal S parameters. Figure 5.1 shows the block diagram of a transistor oscillator. At a certain biasing point, the transistor can be represented by its small-signal S parameters, which are all frequency-dependent variables. For a transistor, the necessary oscillation conditions are given as follows.

$$K = \frac{1 - |S_{11}|^2 - |S_{22}|^2 + |\Delta|^2}{2|S_{12}||S_{21}|} < 1 \tag{5.1}$$

$$\Gamma_{IN}\Gamma_S = 1 \quad \text{and} \quad \Gamma_{OUT}\Gamma_L = 1 \tag{5.2}$$

where

$$\Gamma_{IN} = \frac{Z_{IN} - Z_0}{Z_{IN} + Z_0} = S_{11} + \frac{S_{12}S_{21}\Gamma_L}{1 - S_{22}\Gamma_L} \tag{5.3.1}$$

$$\Gamma_{OUT} = \frac{Z_{OUT} - Z_0}{Z_{OUT} + Z} = S_{22} + \frac{S_{12}S_{21}\Gamma_S}{1 - S_{11}\Gamma_S} \tag{5.3.2}$$

$$\Delta = S_{11}S_{22} - S_{12}S_{21} \tag{5.3.3}$$

Elaborating $\Gamma_{IN}\Gamma_S$ in (5.2), we get

$$\Gamma_{IN}\Gamma_S = \frac{Z_{IN} - Z_0}{Z_{IN} + Z_0} \frac{Z_S - Z_0}{Z_S + Z_0} = \frac{(R_{IN} + jX_{IN}) - Z_0}{(R_{IN} + jX_{IN}) + Z_0} \frac{(R_S + jX_S) - Z_0}{(R_S + jX_S) + Z_0} \tag{5.4}$$

As can be seen from (5.4), the only conditions that make $\Gamma_{IN}\Gamma_S = 1$ are $R_{IN} = -R_S$ and $X_{IN} = -X_S$, leading to the following alternative conditions for oscillation:

$$R_{IN} + R_S = 0 \quad \text{and} \quad X_{IN} + X_S = 0 \tag{5.5}$$

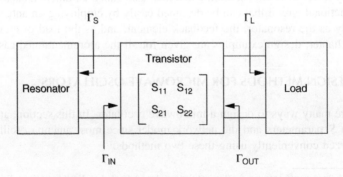

FIGURE 5.1   Block diagram of a transistor oscillator.

Also, it can be shown that the condition $\Gamma_{OUT}\Gamma_L = 1$ in (5.2) is equivalent to the following oscillating conditions:

$$R_{OUT} + R_L = 0 \quad \text{and} \quad X_{OUT} + X_L = 0 \tag{5.6}$$

In brief, the oscillation conditions (5.1) and (5.2) imply (5.5) and (5.6). A complete description of the oscillator theory is available [7].

5.2.2 Design Using a Network Model

Amplifiers require a transistor to have $K > 1$, which is exactly opposite to the requirement for an oscillator. From theory, the same transistor can be used for oscillator design if certain instability is introduced to the amplifier, making $K < 1$. This can be done easily by introducing a positive feedback network to an amplifier. Traditionally, this type of oscillating circuit can best be described and analyzed using network theory [5,6]. Figure 5.2 shows the network representation of an oscillator circuit. Assuming that the amplifier has a gain of $A(\omega)$ and the feedback element provides a positive gain of $B(\omega)$, the transfer function of the network can be written

$$\frac{V_o}{V_i} = \frac{A(\omega)}{1 - B(\omega)A(\omega)} \tag{5.7}$$

For oscillation to occur, the output V_o must exist even when no input signal V_i is applied. This is possible only when $1 - B(\omega)A(\omega) = 0$, leading to the well-known Barkhausen criterion $B(\omega)A(\omega) = 1$. It implies that the loop gain must be unity for oscillation to occur. To make $B(\omega)A(\omega) = 1$, the phase shift around the closed loop must be $0°$ or in multiples of $360°$ at the oscillating frequency.

5.2.3 Specifications of Microwave Oscillators

An oscillator is usually named for the type of resonator it employs. Cavity, dielectrics, microstrip, and yittrium–iron–garnet (YIG) are among the popular

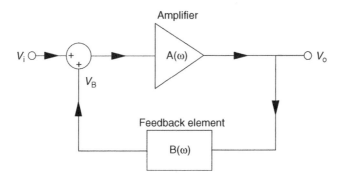

FIGURE 5.2 Network model of an oscillator.

resonators used by oscillators. In general, the design requirements for a microwave oscillator are as follows:

1. High output power
2. Broad frequency tuning range
3. High dc-to-RF conversion efficiency
4. Stable frequency
5. Low phase noise
6. Low power consumption
7. Pure frequency component with low second harmonics

Table 5.1 shows some contemporary oscillators, along with their reported power consumption, output power, and phase noise performance. To design a good microwave oscillator, a high Q factor is always needed for lower phase noise. This can be achieved by incorporating a high-Q resonator and, of course, the use of other high-Q components. Phase-locked loops can be used to improve the oscillator phase noise. It can be seen from Table 5.1 that the phase noise of a phased-locked DRO is higher than that for a free-running DRO. Microstrip oscillators do not have low phase noise, as the Q factor of a microstrip resonator is usually low. Nowadays, having low power consumption and compact size are also important to the design of microwave oscillators. Semiconductor oscillators usually have low power consumption as well as low output power.

TABLE 5.1 Performance of Some Contemporary Microwave Oscillators

Oscillator Name	Operating Frequency (GHz)	Power Consumption (mW)	Output Power (dBm)	Phase Noise (dBc/Hz)
Dielectric resonator oscillator	Phase-locked at 9 [8]	400 [8]	—	−160 at 1 kHz offset [8]
	Free running at 10 [9]	—	12 [9]	−135 at 10 kHz offset [9]
Compact microstrip resonant cell oscillator [10]	Free running at 2.5	140	14.7	−64.7 at 100 kHz offset
Microstrip ring oscillator [11]	51	—	−4	−100 at 1 MHz offset
Semiconductor active voltage-controlled oscillator	Free running at 5.6 [12]	2.4 [12]	−10 [12]	−119.13 at 1 MHz offset [12]
	Free running at 4.8 [13]	13.5 [13]	—	−123.4 at 1 MHz offset [13]
YIG oscillator [4]	Free running at 6	—	14	−130 at 100 kHz offset

5.3 RECENT DEVELOPMENTS AND ISSUES OF ANTENNA OSCILLATORS

Most of the antenna oscillators reported make simultaneous use of the antenna as a resonator, a load, or a feedback element. In most cases, the antenna itself has double functions. When an antenna is used as a resonator or a load, the integrated circuit is called a *reflection-amplifier antenna oscillator*. Active devices such as the BJT and FET have been broadly deployed for such designs. For ease of explanation, a BJT is used to describe all subsequent design examples. As can be seen in Figs. 5.3 and 5.4, a transistor can be configured into either a common-base (or common-gate) or common-emitter (or common-source) biasing scheme in such an antenna oscillator. For such designs, S-parameter analysis is the most intuitive approach. On the other hand, if the antenna is used as a feedback element, as shown in Fig. 5.5, the new device is then called a *coupled-load antenna oscillator*. In this case it can easily be designed and described using the network model.

A good antenna oscillator should have low phase noise, high transmitting power, and broad tunable frequency. The phase noise performance of an antenna oscillator is determined by its Q factor, which is inversely proportional to the total loss of the antenna and oscillator circuit. Antenna oscillators usually have high phase noise. This is because the total Q factor of an antenna oscillator is limited by its radiating antenna, whose Q factor is usually low, a condition necessary for high antenna radiation efficiency and a broad impedance bandwidth. Obviously,

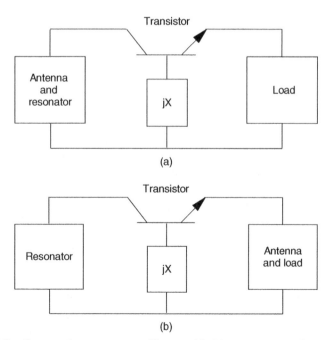

FIGURE 5.3 Common-base antenna oscillators with (a) an antenna as the oscillator resonator; (b) an antenna as the oscillator load.

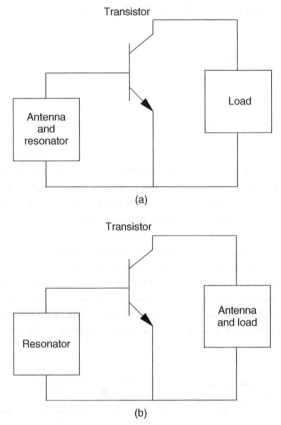

FIGURE 5.4   Common-emitter antenna oscillators with (a) an antenna as the oscillator resonator; (b) an antenna as the oscillator load.

the design of an oscillating antenna is always a trade-off between optimizing the oscillator phase noise, antenna radiation efficiency, and antenna bandwidth. The transmitting power of an antenna oscillator is determined by the output power, linearity, and impedance matching of the active device being employed. It is usually below 25%. Rapid advancement in silicon technology has made possible the use of high-power transistors in designing various antenna oscillators in the millimeter-wave ranges. Like oscillators, oscillating antennas should have a wide frequency tuning range. To achieve this, a tunable voltage-controlled varactor can be used to change the operating frequency of an antenna oscillator.

In Table 5.2 we compare the performances of some recently reported antenna oscillators. Comparing it with Table 5.1, it is obvious that the phase noise of antenna oscillators is higher than for isolated oscillators. Over the past few years, much effort has been made to reduce the phase noise of antenna oscillators. For example, the phase noise of an antenna oscillator can be improved by using an embedded (−101.5 dBc/Hz at 100 kHz offset [17]) or a backing (−96 dBc/Hz

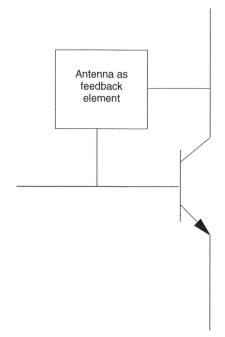

FIGURE 5.5   Configuration of the coupled-load antenna oscillator where the antenna acts as a feedback load.

TABLE 5.2   Performance of Some Recently Reported Antenna Oscillators

Type of Antenna Oscillator	Dc-to-RF Conversion Efficiency	Frequency Tuning Range	Transmitting Power (dBm)	Phase Noise (dBc/Hz)
Substrate integrated waveguide antenna oscillator [14]	6 (%)	1.6 (%)	−2.1	−101 at 1 MHz offset
Microstrip patch antenna oscillator [15]	10.5	—	11.2 (EIRP)	−87.5 at 100 kHz offset
Slot antenna oscillator [16]	—	0.76	8.8	−80 at 100 kHz offset
Solid dielectric resonator antenna oscillator [17]	12.98	—	16.4	−65 at 100 kHz offset

at 100 kHz offset [18,19]) cavity. It has also been found that low phase noise could be obtained by phase- or injection-locking the oscillating antennas (−98 dBc/Hz at 10 kHz offset [20]), at the expense of having a more complex circuit configuration.

5.4   REFLECTION-AMPLIFIER ANTENNA OSCILLATORS

In this section, reflection-amplifier oscillating antennas are discussed. In the first part, a BJT oscillator is combined with a DRA for designing a dielectric resonator antenna oscillator (DRAO). A differential quasi-Yagi antenna oscillator is then explored in the second part. For both cases, the antenna serves as both the radiating element and oscillator load. Finally, a trifunction DRAO, which has an embedded metallic cavity and a better phase noise, is discussed.

5.4.1   Rectangular DRAO

Configuration   A reflection-amplifier DRAO is designed using S parameters. Here, a DR is used simultaneously as the antenna and the oscillator load [17]. The schematic diagram of the common-base and one-port antenna oscillator configuration is shown in Fig. 5.6, where an Infineon BFP420 transistor is used as the active device. As discussed in Section 5.2.1, the circuit oscillates at a particular frequency at which $X_{in} + X_L = 0$ and $|R_{in}| > R_L$. A solid rectangular DRA is used. With reference to Fig. 5.7, the dimensions a, b, and d of the rectangular DRA (with dielectric constant of ε_r) can be determined using the transcendental equation, exciting in its TE_{111}^y mode [21]:

$$k_z \tan \frac{k_z d}{2} = \sqrt{(\varepsilon_r - k_0^2) - k_z^2} \qquad (5.8)$$

where

$$k_x^2 + k_y^2 + k_z^2 = \varepsilon_r k_0^2 \qquad (5.9.1)$$

$$k_x = \frac{\pi}{a} \qquad k_y = \frac{\pi}{b} \qquad (5.9.2)$$

k_x, k_y, and k_z represent the wavenumbers in the x-, y-, and z-directions, respectively and k_0 is the free-space wavenumber at resonance frequency.

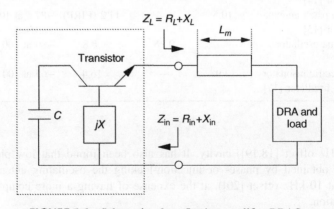

FIGURE 5.6   Schematic of a reflection-amplifier DRAO.

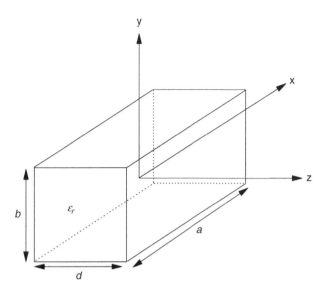

FIGURE 5.7   Dimensions of an isolated rectangular solid DRA.

This DRA is then used in designing the DRAO. Figure 5.8 shows the configuration of the DRAO. The DR has a length of $L = 52.2$ mm, a width of $W = 42.4$ mm, a height of $H = 26.1$ mm, and a dielectric constant of $\varepsilon_r = 6$. It is excited by a microstrip-fed aperture of length $L_a = 0.3561\lambda_e$ and $W_a = 2$ mm, where $\lambda_e = \lambda_o/\sqrt{(\varepsilon_r + \varepsilon_{rs})/2}$ is the effective wavelength in the aperture. Duroid substrates of dielectric constant $\varepsilon_r = 2.94$ and $d = 0.762$ mm were used for all the circuits.

As can be seen in Fig. 5.8, the emitter output of the transistor was soldered to the 50-Ω microstrip feedline which was used to feed the antenna. The feedline was extended with a length of $L_m = 40$ mm and a matching stub of $L_s = 9.5$ mm was used. A finite ground plane of size 20×20 cm^2 was used in both the simulations and experiments for the DRAO. In the simulation and optimization processes, the DRA in Fig. 5.6 was replaced by a 50-Ω resistive load. After optimization, the resistor was then replaced by the actual antenna. This is feasible because the input impedance of the DRA can be approximated by a resistive load (50 Ω in this case) at resonance.

Measurement Method   In this section we explore the use of the compact-range anechoic chamber for measuring active antennas. Here, the HP85301CK02 receiving-mode compact-range antenna measurement system [22] was used to measure the DRAO, in the arrangement shown in Fig. 5.9. The distances are $D_1 = 4.5$ m and $D_2 = 2.5$ m to ensure that the far-field requirement is satisfied at 1.85 GHz, where the side-feed receiving horn antenna has an antenna gain of $G_r = 7.36$ dBi. To calibrate the measurement system, the passive rectangular solid DRA was first placed at the center of the rotator. It was then supplied with

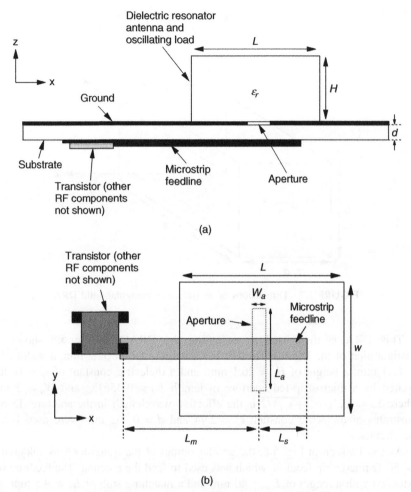

FIGURE 5.8 Rectangular solid DRAO excited by a microstripline-fed aperture: (a) side view; (b) top view. (From [17], copyright © 2007 IEEE, with permission.)

a 1.85-GHz monotonic signal with various power levels. In the measurements, the monotonic testing signal was generated by an Agilent E8244A signal generator, and the received power was recorded at the receiving end. Figure 5.10 shows the curve that relates the transmitted and received powers of the passive rectangular solid DRA. The curve of a passive hollow DRA (which is supplied with a 1.90-GHz monotonic signal and will be used for designing the trifunction DRA later) is also given in the figure. The calibration has taken into account all of the gains and losses, including the horn gain, DRA gain, free-space loss, and other losses, as depicted in the inset. Since the output power of the signal generator is known for the passive DRA, the DRAO output power can be determined by comparing the received power from the DRAO directly to that from the passive DRA.

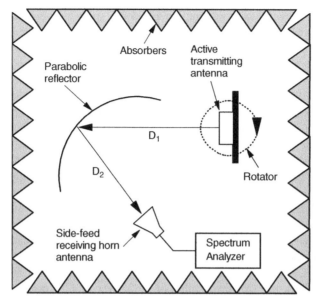

FIGURE 5.9   Modified compact-range measurement setup for active transmitting antennas. (From [17], copyright © 2007 IEEE, with permission.)

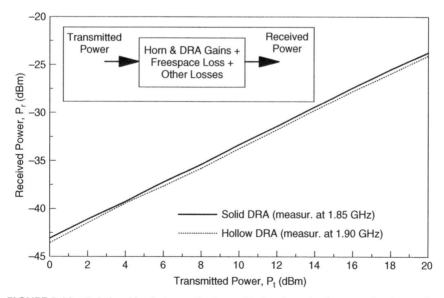

FIGURE 5.10   Relationships between the transmitted and received powers for the passive solid DRA in Fig. 5.8 and the passive hollow DRA in Fig. 5.15. (From [17], copyright © 2007 IEEE, with permission.)

Results and Discussion   The antenna part of the DRAO was measured using an HP8510C network analyzer, and an HP8595E spectrum analyzer was used to characterize the oscillation performance of the DRAO. In experiment, the BFP420 transistor is biased at $V_{CE} = 2.2$ V and $I_C = 27$ mA, with biasing resistors of $R_1 = 390$ Ω and $R_2 = 47$ Ω. The dc biasing voltages are $V_1 = 12.42$ V and $V_2 = -2$ V. A dc block and two RF chokes are used. Before being connected to the DRA, the oscillator was optimized experimentally with a 50-Ω resistor. An inductive load of $X = j59.68$ Ω ($L = 5.1$ nH) was used to destabilize the transistor, with a capacitor of $C = 6$ pF terminated at the collector port. The Eccostock HiK Powder with dielectric constant $\varepsilon_r = 6$ was used as the dielectric material. A few hard-form clad boards ($\varepsilon_r \sim 1$) were used to construct a rectangular container for the powder, which is fine granular and free flowing. It was found that slight jogging is sufficient to even the powder in the container and no further processing is required.

Ansoft HFSS software was used in the following simulations. Figure 5.11 shows the simulated and measured reflection coefficients shown as a function of frequency for the passive solid DRA in Fig. 5.8. The corresponding impedances are given in the inset. As can be seen from the figure, a wide measured impedance bandwidth of 22.14% is obtained. The measured resonance frequency is 1.86 GHz, which agrees well with the simulated value of 1.83 GHz (1.47% error). By using the compact-range measurement method described in this section, the measured spectrum of the free-running active DRAO is shown in Fig. 5.12. With reference to the figure, a stable oscillation of a received power level of $P_r = -27.04$ dBm is observed at 1.878 GHz with a clean spectrum. Using Fig. 5.10, the transmitting

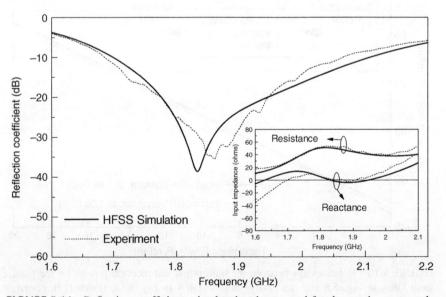

FIGURE 5.11   Reflection coefficients simulated and measured for the passive rectangular solid DRA in Fig. 5.8. The inset shows the input impedances simulated and measured as a function of frequency. (From [17], copyright © 2007 IEEE, with permission.)

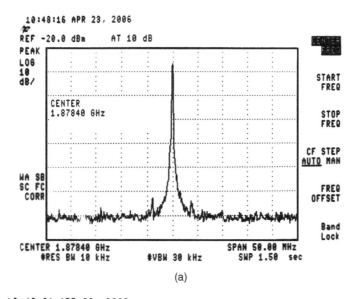

(a)

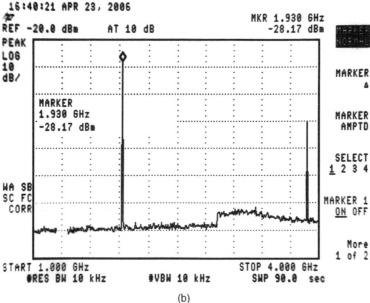

(b)

FIGURE 5.12   Power spectrum shown as a function of frequency for the rectangular solid DRAO in Fig. 5.8. The center frequency is 1.8784 GHz with a span of 50 MHz. (a) Fundamental oscillating signal; (b) fundamental oscillating signal and second harmonic. (From [17], copyright © 2007 IEEE, with permission.)

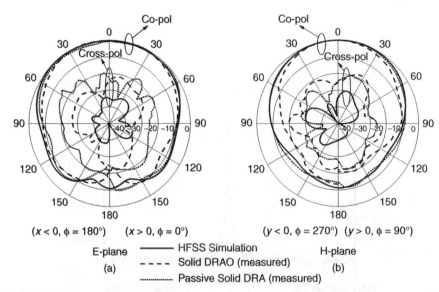

E-plane ——— HFSS Simulation H-plane
(a) – – – – Solid DRAO (measured) (b)
·············· Passive Solid DRA (measured)

FIGURE 5.13 Normalized radiation patterns simulated and measured for the rectangular solid DRAO in Fig. 5.8: (a) E-plane (xz-plane). (b) H-plane (yz-plane). The frequencies of the simulated and measured results are 1.83 and 1.878 GHz, respectively. (From [17], copyright © 2007 IEEE, with permission.)

power of the DRAO is estimated to be $P_t = 16.4$ dBm. The dc–RF conversion efficiency is 12.98%. It has phase noise of about -103 dBc/Hz at 5 MHz offset (about -65 dBc/Hz at 100 kHz offset). As can be seen from Fig. 5.12(b), the second harmonic is 22.02 dB lower than the fundamental oscillating signal. The normalized field patterns of the active DRAO and the passive rectangular solid DRA were measured and shown in Fig. 5.13. For ease of reference, the simulated field patterns are also shown in the same figure. Good agreement is observed between the simulations and measurements. The co-polarized fields are generally 20 dB stronger than the cross-polarized fields in the bore-sight direction, showing that the antenna part of the DRAO is a good linearly polarized EM radiator. In Fig. 5.13, the measured cross-polarized fields are observed to be 10 to 20 dB higher than the simulated ones, which should be due to imperfections of the experiment. Figure 5.14 shows the measured antenna gain. The DRA has an antenna gain ranging from 3 to 7 dBi around resonance.

5.4.2 Hollow DRAO

Configuration Figure 5.15 shows the configuration of a trifunction hollow DRAO [17]. Now, the DR functions simultaneously as an antenna, a packaging cover, and an oscillator load. To design the DRAO, a notched DR is first formed by removing the lower central portion of the solid DR (Fig. 5.8). With the use of the design procedure mentioned in Section 2.4.1, the hollow DRA is then constructed by

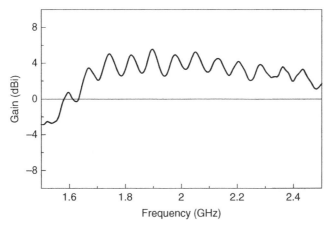

FIGURE 5.14   Antenna gain measured for the rectangular DRA in Fig. 5.8. (From [17], copyright © 2007 IEEE, with permission.)

covering up the two sides which are parallel to the xz-plane of the notched DR. The DRA was designed to resonate fundamentally at 1.85 GHz. The optimized dimensions of the DR were found to be $L = 52.2$ mm, $W = 56.4$ mm, $H = 26.1$ mm, $a = 26$ mm, $b = 42.4$ mm, and $h = 9$ mm. As can be seen in Fig. 5.15, the inner surface of the DR has a conducting coating to isolate the DR from the active circuits placed inside the hollow region. The ground plane of the substrate and the four metallic supports form a rectangular metallic cavity. On top of the substrate, a microstrip-fed aperture of $L_a = 0.434\lambda_e$, $W_a = 2$ mm was used to excite the DRA. A shorter interconnecting stub with a length of $L_m = 10$ mm was used. The length of the matching stub is given by $L_s = 7.5$ mm. In an experiment a hole was drilled in the ground plane to supply dc bias to the oscillator circuit. An aluminum ground plane of 30 (x-direction) × 20 cm (y-direction) in size was used in the experiment. Since the computer memory is limited, a smaller ground plane of 20 × 20 cm was used in HFSS simulations. The same oscillator (as the rectangular solid DRAO) and HiK powder ($\varepsilon_r = 6$) were used to construct the hollow DRAO.

Results and Discussion   Figure 5.16 shows the reflection coefficients simulated and measured for the passive hollow DRA (Fig. 5.15), where reasonable agreement is observed. The corresponding input impedances are also shown in the inset. The impedance bandwidth measured ($|S_{11}| < -10$ dB) is 5.11%. With reference to the figure, the resonance frequencies measured and simulated are 1.853 and 1.838 GHz, respectively, with an error of 0.816%. Again, with the use of the experimental setup in Fig. 5.9, the power spectrum measured for the trifunction hollow DRAO is illustrated in Fig. 5.17. A stable oscillating frequency of about 1.90 GHz with a clean spectrum is observed. With reference to the figure, the power received, $P_r = -26.5$ dBm, was measured at the horn by a spectrum analyzer, which gives an estimated transmitting power of $P_t = 17.2$ dBm by referring to Fig. 5.10. The dc–RF conversion efficiency is about 15.6%. As can be seen from Fig. 5.17(b), it is

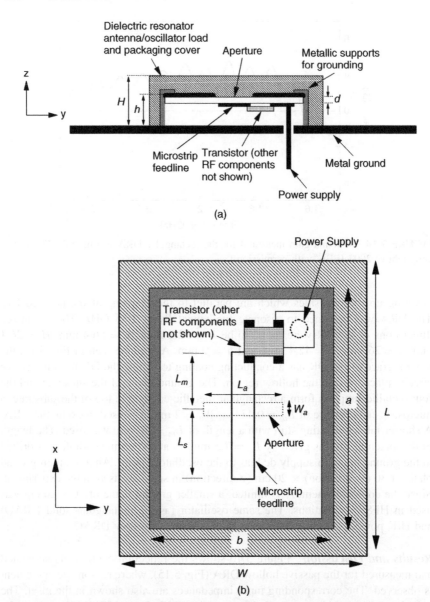

FIGURE 5.15 Rectangular hollow DRAO excited by a microstripline-feed aperture: (a) front view; (b) top view. (From [17], copyright © 2007 IEEE, with permission.)

noted that the second harmonic is 21.59 dB lower than its fundamental oscillating signal. Comparing with the rectangular solid DRAO, the hollow DRAO in this case has a much better phase noise of −101.5 dBc/Hz at 100 kHz offset because it has a higher loaded Q factor. It has been studied that the Q factor of a microstrip antenna oscillator can be increased by loading an external rectangular metallic

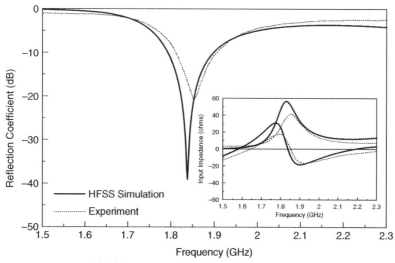

FIGURE 5.16 Reflection coefficients simulated and measured for the rectangular hollow DRA in Fig. 5.15. The inset shows the input impedances simulated and measured as a function of frequency. (From [17], copyright © 2007 IEEE, with permission.)

cavity on the reverse side of the antenna [18,19]. Obviously, the hollow DRAO in Fig. 5.15 has a much more compact size since the metallic cavity is embedded inside the DRA, wasting no extra space. This compact microwave structure provides a possible solution for the IC packaging. Using the measurement setup in Fig. 5.9, the normalized radiation patterns simulated and measured for the passive DRA and active DRAO are shown in Fig. 5.18. Again, the cross-polarized fields are generally 20 dB weaker than their co-polarized counterparts in both E- (xz-plane) and H-plane (yz-plane). As can be seen from Fig. 5.19, the measured antenna gain fluctuates in the range 3 to 6 dBi in the antenna passband.

5.4.3 Differential Planar Antenna Oscillator

Configuration Figure 5.20 shows the configuration of a differential antenna oscillator [23]. Here, a quasi-Yagi antenna is used as the oscillator load of a voltage-controlled integrated antenna oscillator (VCIAO). The antenna part of the VCIAO is discussed first. The top and side views of the quasi-Yagi–Uda antenna are shown in Fig. 5.21. The antenna is made on a three-layer substrate with dielectric constant $\varepsilon_r = 3.48$ and thickness $h = 1.016$ mm. As can be seen from the figure, the middle layer contains a director and an inserted conductor plane. The conductive middle layer is also a virtual ground plane and a reflector of the antenna. A feeding 100-Ω double-sided parallel-strip line (DSPSL) (with a linewidth of W_{fed}) is made up of the metallic lines on the top and bottom layers. As the DSPSL itself is a good balanced transmission line, the signals on the top and bottom lines are constantly frequency independent out-of-phase. It eliminates the need of any additional balun for feeding the antenna. The optimized design parameters of the quasi-Yagi are given by $W_{\text{fed}} = 1.26$ mm, $L_{\text{gnd}} = 14$ mm, $L_{\text{dri}} = 28.63$ mm, $L_{\text{dir}} =$

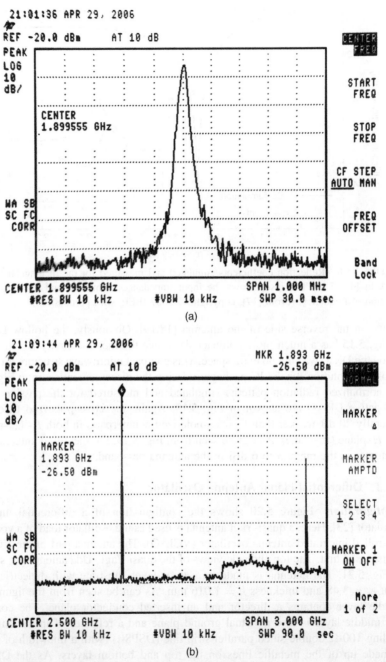

FIGURE 5.17 Power spectrum as a function of frequency for the hollow DRAO in Fig. 5.15. The center frequency is 1.899555 GHz with a span of 1 MHz. (a) Fundamental oscillating signal; (b) fundamental oscillating signal and second harmonic. (From [17], copyright © 2007 IEEE, with permission.)

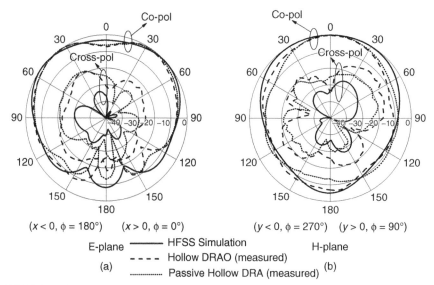

FIGURE 5.18 Normalized radiation patterns simulated and measured for the rectangular hollow DRAO in Fig. 5.15: (a) E-plane (xz-plane); (b) H-plane (yz-plane). The frequencies of the simulated and measured results are 1.838 and 1.90 GHz, respectively. (From [17], copyright © 2007 IEEE, with permission.)

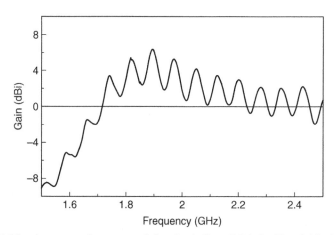

FIGURE 5.19 Antenna gain measured for the hollow DRA in Fig. 5.15. (From [17], copyright © 2007 IEEE, with permission.)

35 mm, $S_{\text{fed}} = 21.5$ mm, $S_{\text{dir}} = 5.1$ mm, $W_{\text{dri}} = W_{\text{dir}} = 5$ mm, $a = 65$ mm, and $b = 70$ mm.

HFSS was used to simulate the antenna part of the VCIAO. The reflection coefficient simulated for the passive antenna (shown in Fig. 5.21) is shown in Fig. 5.22. As can be seen from the figure, the impedance bandwidth simulated ($|S_{11}| < -10$ dB) is 34%. With reference to the inset, the input resistance is in

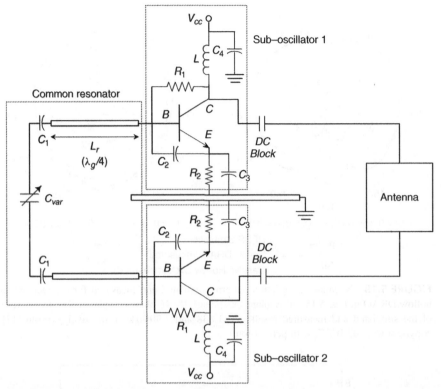

FIGURE 5.20 Schematic of a differential quasi-Yagi antenna oscillator. (Courtesy of J. Shi, J. X. Chen, and Q. Xue, City University of Hong Kong. From [23], copyright © 2007 IEEE, with permission.)

the range 90 ± 10 Ω and the reactance is in the range 0 ± 15 Ω. The gain is 5.5 to 6.5 dBi across the entire passband in the frequency range 2.2 to 2.8 GHz. The radiation patterns simulated at 2.7 GHz are shown in Fig. 5.23. In both the E- and H-planes, the radiation patterns are broad-side, with a front-to-back ratio better than 17 dB.

Next, a differential voltage-controlled integrated antenna oscillator (VCIAO) is constructed by inserting a pair of Infineon BFP650 transistors onto the top and bottom layers, right above and below the inserted conductor planes, of the substrate. The configuration is shown in Fig. 5.24. For brevity, the biasing and feedback circuits (dc voltage supply V_{cc}, capacitors C_2 to C_4, resistors R_1 to R_2, and inductor L in Fig. 5.20) are not shown. The middle-layer conductor plane serves as the ground for the differential VCO as well as the reflector for the antenna. As can be seen from Fig. 5.24(c), a DSPSL-fed quasi-Yagi antenna is built on the right-hand side of the inserted ground plane while a quarter-wavelength resonator is built on the left-hand side. A varactor was embedded in a perforation penetrating the substrate, and it was connected to the $\lambda_g/4$ resonator going through a capacitor C_1. Obviously, the differential VCO is composed by two identical suboscillators

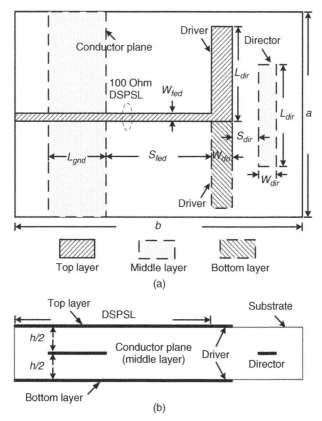

FIGURE 5.21 Layout of a DSPSL-fed quasi-Yagi antenna: (a) top view; (b) side view. (Courtesy of J. Shi, J. X. Chen, and Q. Xue, City University of Hong Kong. From [23], copyright © 2007 IEEE, with permission.)

being placed on the two opposite surfaces of the substrate. The design parameters of the VCIAO are as follows: $L_r = 15.8$ mm, $W_r = 0.5$ mm, $a = 65$ mm, $b = 80$ mm, and the dimension of the antenna is identical to that of the passive case (Fig. 5.21). The chip capacitor C_1 is 1 pF and the varactor diode is Skyworks' SMV1247-079LF. Both transistors are biased at $V_{CE} = 3.5$ V with a total current of $I_C = 41$ mA. A fabricated VCIAO is shown in Fig. 5.25.

As mentioned by Shi et al. [23], the co-design procedure of the VCIAO can be summarized as follows:

1. An isolated quasi-Yagi antenna (with 100-Ω DSPSL feedline) is first simulated and optimized. Next, the input impedance of the antenna is extracted.
2. The value of the antenna input impedance is then used as the oscillator load of a differential VCO. The complete circuit is called VCIAO, and it is optimized again to oscillate at a certain frequency. In this case, complex conjugate match is desired for maximum radiation power.

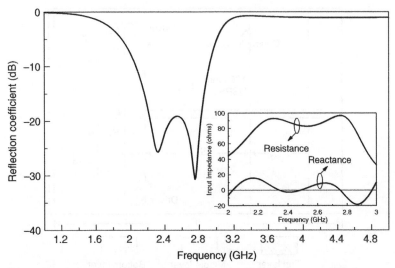

FIGURE 5.22 Reflection coefficient simulated for the passive quasi-Yagi antenna in Fig. 5.21. The inset shows the input impedance simulated as a function of frequency. (Courtesy of J. Shi, J. X. Chen, and Q. Xue, City University of Hong Kong. From [23], copyright © 2007 IEEE, with permission.)

3. Fine-tuning the operating frequency of the antenna oscillator by changing the length L_r of the $\lambda_g/4$ resonator. The values of the chip capacitor C_1 and varactor C_{var} can also be used for tuning.

Measurement Method The transmitting power from the VCIAO is measured using the active antenna test range setup [1] in an anechoic chamber. Use is made of the Friis transmission equation for calculating the radiated power of the antenna oscillator. The measurement setup is shown in Fig. 5.26. Different from the measurement setup in Section 5.4.1, here the radiated power (P_r) of the active transmitting antenna is directly illuminated on the receiving reference horn. A short coaxial cable (cable loss of $L_{cable} = -2$ dB) is used to connect the double-ridged horn antenna (antenna gain $G_r = 8.3$ dBi) to the HP 8593E spectrum analyzer for measuring the received power. The separation distance is $D = 4$ m, being in the far field. The effective isotropic radiated power (EIRP) can be calculated using (5.10). The information of the transmitting antenna gain (G_t) is included by EIRP.

$$\text{EIRP} = P_r - L_{cable} - G_r + 10\log\left(\frac{4\pi D}{\lambda_g}\right)^2 \qquad (5.10)$$

When the varactor is biased at 2.7 V, the radiated power is found to be $P_r = -25.6$ dBm at 2.69 GHz (shown in Fig. 5.27). It can easily be calculated from (5.10) that the corresponding EIRP is 21.2 dBm.

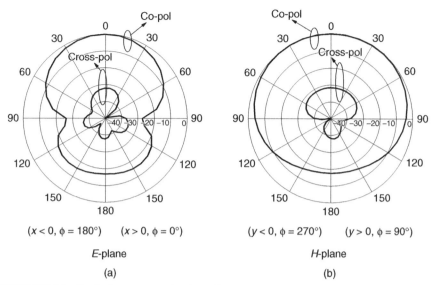

FIGURE 5.23   Radiation patterns simulated for the passive quasi-Yagi antenna in Fig. 5.21. (a) E-plane; (b) H-plane. (Courtesy of J. Shi, J. X. Chen, and Q. Xue, City University of Hong Kong. From [23], copyright © 2007 IEEE, with permission.)

Results and Discussion   In Fig. 5.28, the phase noise of the VCIAO at the oscillation frequency of 2.69 GHz is shown. With reference to the figure, the phase noise reads about -92 dBc/Hz at 100 kHz offset and -112 dBc/Hz at 500 kHz offset. As can be seen from Fig. 5.29, the tuning range of the VCIAO is 2.585 to 2.785 GHz, with about 1.4 dB of amplitude variation. From the figure it can be observed that the antenna oscillator has a gain of 66.7 MHz/V in the linear tunable range 2.61 to 2.73 GHz.

By rotating the rotator, the co- and cross-polarized radiation patterns were measured in the E- and H-planes at oscillation frequencies of 2.61, 2.69, and 2.77 GHz. The results are shown in Fig. 5.30. At all three frequencies, the co-polarized fields are greater than their cross-polarized counterparts by at least 20 dB. This shows that the VCIAO has a good linear polarization. In general, the front-to-back ratio is better than 16 dB.

5.5   COUPLED-LOAD ANTENNA OSCILLATORS

5.5.1   Coupled-Load Microstrip Patch Oscillator

Configuration   By making an antenna simultaneously the feedback load of an amplifier, the coupled-load antenna oscillator can easily be designed, as noted in Section 5.2.2 [24]. To demonstrate the design idea, the Mini-Circuits ERA-2SM broadband RF amplifier is integrated with a microstrip dual-patch antenna for the construction of a coupled-load antenna oscillator [25]. The schematic of the antenna

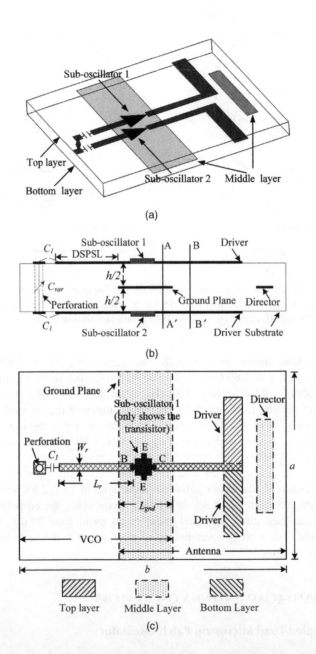

FIGURE 5.24 Configuration of a VCIAO: (a) three-dimensional view; (b) side view; (c) top view. (Courtesy of J. Shi, J. X. Chen, and Q. Xue, City University of Hong Kong. From [23], copyright © 2007 IEEE, with permission.)

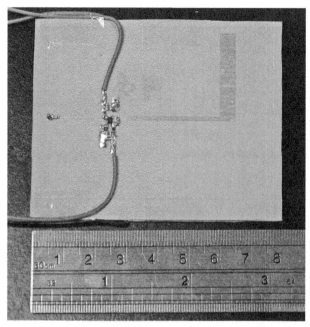

FIGURE 5.25   Top view of the VCIAO in Fig. 5.24. (Courtesy of J. Shi, J. X. Chen, and Q. Xue, City University of Hong Kong. From 5 [23], copyright © 2007 IEEE, with permission.)

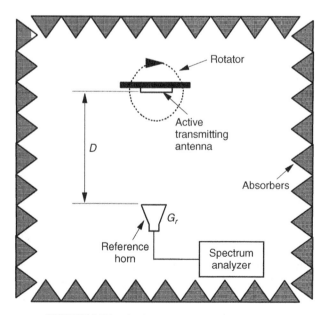

FIGURE 5.26   Active antenna test range setup.

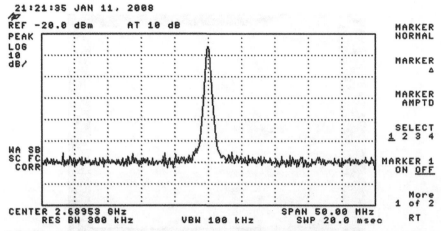

FIGURE 5.27 Output power spectrum received at a frequency of 2.69 GHz for the VCIAO in Fig. 5.24. (Courtesy of J. Shi, J. X. Chen, and Q. Xue, City University of Hong Kong. From [23], copyright © 2007 IEEE, with permission.)

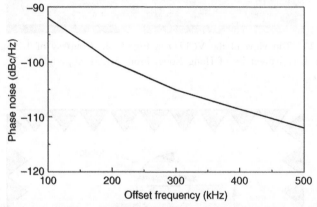

FIGURE 5.28 Phase noise measured for the VCIAO in Fig. 5.24. (Courtesy of J. Shi, J. X. Chen, and Q. Xue, City University of Hong Kong. From [23], copyright © 2007 IEEE, with permission.)

oscillator is shown in Fig. 5.31, where the numberings represent the chip pinouts. The design parameters are $R_b = 100\ \Omega$, $C_{block} = 1000$ pF, RFC = 680 nH, and $V_{cc} = 7.5$ V. It is given in the datasheet that the amplifier has a typical gain of about 16 dB at 1 GHz. With reference to Fig. 5.32(a), the antenna is designed from a pair of symmetric rectangular ($\sim \lambda/2$) patches with corner cutouts for accommodating the amplifier. It is designed on a FR-4 (with a thickness of 62 mils) with dimension $D_1 = 48.5$ mm and $D_2 = 59$ mm. The prototype is shown in Fig. 5.32(b). To ensure stable oscillation, according to the Barkhausen criterion it is imperative that $|H_{ant}(j\omega)H_{amp}(j\omega)| = 1$ and $\angle H_{ant}(j\omega)H_{amp}(j\omega) = 0°$. Here $H_{amp}(j\omega)$ and $H_{ant}(j\omega)$ are the transfer functions of the amplifier and antenna, respectively.

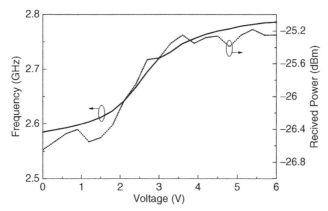

FIGURE 5.29 Results of the tuning frequency range and received output power measured for the VCIAO in Fig. 5.24. (Courtesy of J. Shi, J. X. Chen, and Q. Xue, City University of Hong Kong. From [23], copyright © 2007 IEEE, with permission.)

Results and Discussion Figure 5.33 shows the simulated open-loop magnitude $|H_{ant}(j\omega)H_{amp}(j\omega)|$ and the simulated open-loop angle $\angle H_{ant}(j\omega)H_{amp}(j\omega)$. Oscillation occurs at around the frequency marked by a diamond-shaped marker. At this frequency, the open-loop angle is $0°$. The oscillation frequency is measured by a receiving antenna (gain ~5 dBi) at a distance of 1 m. With reference to Fig. 5.34, the peak power is about -33 dB, with a center oscillation frequency of 1.261 GHz. The phase noise measured for the antenna oscillator is about -120 dBc/Hz at 1 MHz, as shown in Fig. 5.35. By tuning the V_{cc} over a range of 6.5 to 8.1 V, the tuning sensitivity measured for the antenna oscillator is found to be about 1.13 MHz/V, as shown in Fig. 5.36. The radiation fields were measured in the E-plane and the result is shown in Fig. 5.37. In general, the co-polarized field is isolated from its cross-polarized counterpart by at least 20 dB.

5.5.2 Patch Antenna Oscillator with Feedback Loop

Configuration A very compact feedback-type antenna oscillator was designed by Mueller et al. [26] on a high-dielectric-constant microwave substrate ($\varepsilon_r = 10.2$ and thickness-0.635 mm). In this case, a capacitor-loaded microstrip antenna is used as the feedback element of an oscillator. Despite being placed on a high-permittivity substrate, the radiation efficiency of the antenna oscillator remains good, achieving a high radiation power. The configuration of the antenna oscillator is shown in Fig. 5.38. The antenna consists of two patches interconnected by a narrow stripline. A capacitor is inserted into the middle point of the narrow line to construct a feedback loop. The antenna, capacitor, and feedlines have a total electrical length ($\angle S_{21antenna}$) of $\frac{3}{2}\lambda_g$, where λ_g is the guided wavelength. It was found that the antenna has current distribution similar to that for a conventional microstrip patch antenna. To oscillate, the transistor has to provide an additional phase ($\angle S_{21transistor}$) so that $\angle S_{21antenna} + S_{21transistor} = 0°$. The NEC super-low-noise high-frequency field-effect transistor (HF FET) NE3210S01 is used for the oscillator design [27]. This is a pseudomorphic heterojunction field-effect transistor, which has a 0.20-μm gate length and a 160-μm gate width. The transistor was biased using a single 1.5-V battery placed between the source and drain terminals. The gate terminal is left open. The amplifier gain of this transistor is about 8.5 to 9 dB over the frequency range 6 to 12 GHz. A 1.2-pF capacitor was used as the dc isolator

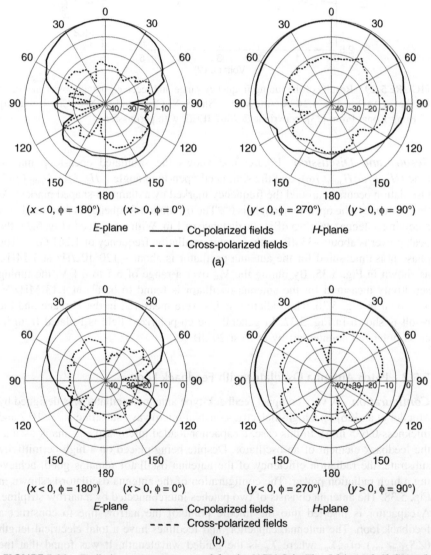

FIGURE 5.30 Radiation patterns measured for the VCIAO in Fig. 5.24: (a) 2.62 GHz; (b) 2.69 GHz; (c) 2.76 GHz. (Courtesy of J. Shi, J. X. Chen, and Q. Xue, City University of Hong Kong. From [23], copyright © 2007 IEEE, with permission.)

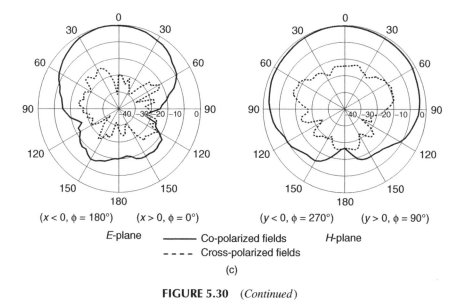

$(x < 0, \phi = 180°)$ $(x > 0, \phi = 0°)$ $(y < 0, \phi = 270°)$ $(y > 0, \phi = 90°)$

E-plane ———— Co-polarized fields *H*-plane
        - - - - Cross-polarized fields

(c)

FIGURE 5.30 (*Continued*)

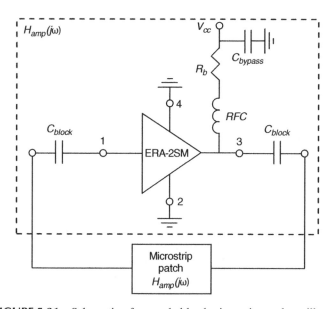

FIGURE 5.31 Schematic of a coupled-load microstrip patch oscillator.

between the drain and gate as well as the feedback element of the antenna oscillator. Both the antenna and oscillator are fabricated on a RT/Duroid 6010LM substrate ($\varepsilon_r = 10.2$ and thickness $= 0.635$ mm). The antenna oscillator is designed to operate at 8.50 GHz, occupying an area of 5×6 mm^2. The active integrated antenna shows stable oscillation and an excellent radiation pattern across the entire X band.

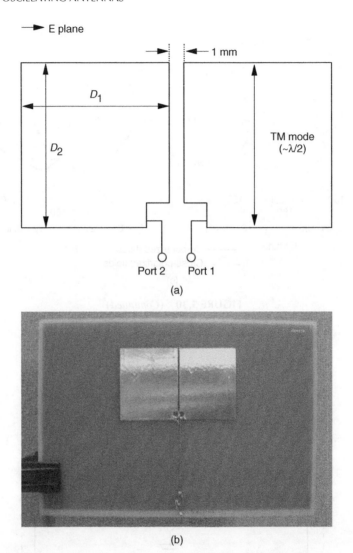

(a)

(b)

FIGURE 5.32 (a) Configuration of a microstrip dual-patch antenna; (b) photograph of a microstrip dual-patch antenna. (Courtesy of S. Makarov, Worcester Polytechnic Institute. From [25], copyright © 2005 IEEE, with permission.)

Results and Discussion A calibrated Agilent E8361A PNA and an Agilent 4448A spectrum analyzer were used to measure the S parameters and power spectrums, respectively. The radiation measurements were performed in a far-field anechoic chamber located at the NASA–Glenn Research Center (Cleveland, OH). A standard horn antenna ($G_r = 10$ dBi at 8.5 GHz) was used. By connecting the horn to the

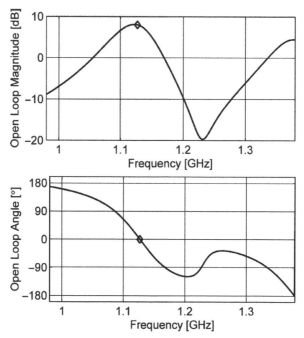

FIGURE 5.33 Open-loop magnitude and angle simulated for the antenna oscillator in Fig. 5.31. (Courtesy of S. Makarov, Worcester Polytechnic Institute.)

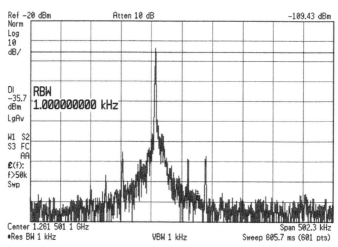

FIGURE 5.34 Power spectrum as a function of frequency for the microstrip dual-patch antenna oscillator in Fig. 5.31. The center frequency is 1.261 GHz with a span (range of x-axis) of 502.3 kHz. (Courtesy of S. Makarov, Worcester Polytechnic Institute.)

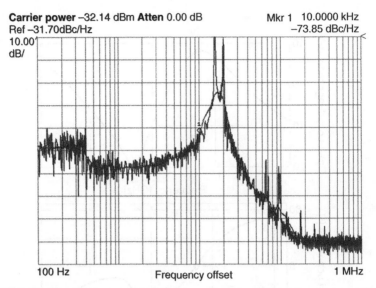

FIGURE 5.35 Phase noise measured for the microstrip dual-patch antenna oscillator in Fig. 5.31. (Courtesy of S. Makarov, Worcester Polytechnic Institute.)

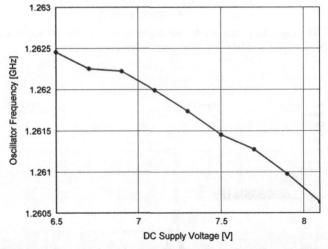

FIGURE 5.36 Tuning sensitivity measured for the microstrip dual-patch antenna oscillator in Fig. 5.31. (Courtesy of S. Makarov, Worcester Polytechnic Institute. From [25], copyright © 2005 IEEE, with permission.)

spectrum analyzer, the power radiated was measured at a distance of 135 cm. For a characterization of the radiation patterns, an HP 437 power meter was used to measure the received powers of the co- and cross-polarized fields, with the horn placed at a distance of 665 cm. In this case, a stepping-motor-driven rotational stage, rotating with an 0.25° increment, was employed.

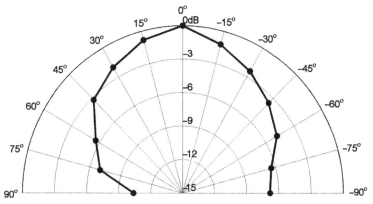

FIGURE 5.37 Co-polarized pattern measured in the E-plane. (Courtesy of S. Makarov of Worcester Polytechnic Institute. From [25], copyright © 2005 IEEE, with permission.)

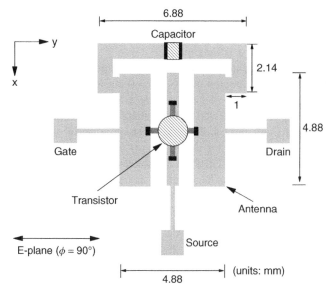

FIGURE 5.38 Configuration of an antenna oscillator with a capacitor-loaded feedback patch antenna.

The frequency discriminator method [28] was adopted to measure phase noise. An Agilent E5500 phase noise measurement system was used. The measurement procedure is described briefly here. The received signal is first split into two paths. A coaxial delay line (35 ns) is inserted into one of the signal paths so that the noise between the two paths is uncorrelated at the offset frequency. The delay line also acts as a frequency-to-phase transformer. Then an amplifier is placed in the same path to compensate the insertion loss of the delay line. A manually tunable phase shifter (HP X885 A) is connected to another signal path so that the signals in the two paths are at quadrature. Finally, the Agilent E5500 is used as a mixer, and the phase noise is measured using an Agilent 4411 B ESA-L series spectrum analyzer. To determine the phase detector constant, an Agilent E8257C PSG analog signal generator is used to perform the double-sided spur calibration.

The test fixture in Fig. 5.39 is deployed to measure concurrently the S parameters and radiated power of the antenna oscillator in an anechoic chamber. The transistor drain (port 1) and gate (port 2) terminals are wire-bonded to two microstrips of the test fixture for connection to a calibrated Agilent E8361A PNA with an input power of -15 dBm. Transmission coefficient (S_{21}) is then taken for three settings: (1) with the transistor removed, (2) with the transistor mounted but no bias ($V_D =$ open) applied, and (3) with the transistor mounted and biased ($V_D = 1.2$ V, with the source grounded and the gate opened). The transmission coefficients measured are shown in Fig. 5.40. As can be seen from the figure, there is a peak observed at 8.65 GHz, with a Q_L value of about 70. With the transistor biased, oscillation occurs as $\angle S_{21\text{antenna}} + S_{21\text{transistor}} = 0°$. In this case, the transmission coefficient is about 19 dB at 8.56 GHz. The inset of Fig. 5.40 shows the concurrently measured radiation power of the antenna oscillator. It shows clearly that the device radiates at the frequency where S_{21} is maximum.

Next, the antenna oscillator is unsoldered from the test fixture. To measure the radiation pattern, it is remounted on a 3.0×3.0 cm flat metal plate. The transistor is biased at $V_g =$ open, $V_d = 1.5$ V, and $I_d = 17.7$ mA. Figures 5.41 and 5.42 show the power spectrum received and the phase noise measured, respectively. At 100 kHz offset, the antenna oscillator has a phase noise of -87.5 dBc/Hz, which is quite high, as the Q factor for the antenna is usually low.

At the receiving horn, the EIRP of the antenna oscillator can be calculated by EIRP $= P_t D_t = P_r/G_r(\lambda_0/4\pi R)$, where P_t is the transmitted power and D_t is the directivity of the antenna oscillator [29]. P_r ($= -46.3$ dBm) is the power received by the horn (G_r), λ_0 ($= 3.54$ cm at 8.48 GHz) is the free-space wavelength, and R ($= 665$ cm) is the distance between the two antennas. The EIRP of the antenna oscillator is then calculated as 11.2 dBm. Assuming that the antenna is

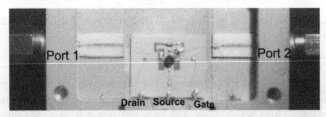

FIGURE 5.39 Antenna oscillator with a capacitor-loaded feedback patch antenna. (Courtesy of F. A. Miranda, NASA–Glenn Research Center, Cleveland, Ohio. From [26], copyright © 2008 IEEE, with permission.)

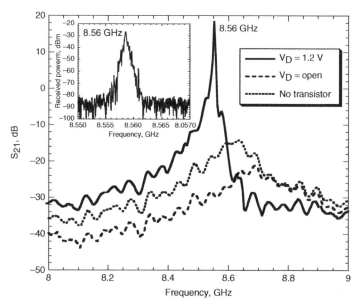

FIGURE 5.40 Transmission (S_{21}) versus frequency for the antenna oscillator under three different settings. The inset shows the measured radiated power, with the detector located 37 cm from the antenna oscillator. (Courtesy of F. A. Miranda, NASA–Glenn Research Center, Cleveland, Ohio. From [26], copyright © 2008 IEEE, with permission.)

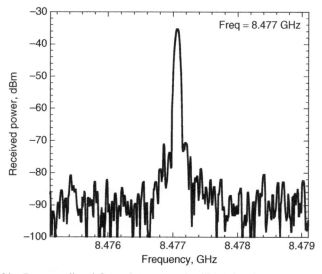

FIGURE 5.41 Power radiated from the antenna oscillator in Fig. 5.38, with the detector located 132 cm away (resolution bandwidth = video bandwidth = 47 kHz). (Courtesy of F. A. Miranda, NASA–Glenn Research Center, Cleveland, Ohio. From [26], copyright © 2008 IEEE, with permission.)

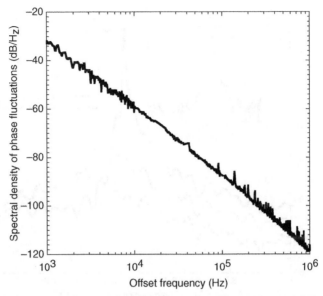

FIGURE 5.42   Phase noise measured for the antenna oscillator ($V_{DS} = 1.5$ V) in Fig. 5.38. (Courtesy of F. A. Miranda, NASA–Glenn Research Center, Cleveland, Ohio. From [26], copyright © 2008 IEEE, with permission.)

unidirectional and the directivity can be estimated by $D_t = 41,253/\theta_1\phi_1 = 6.74$ dB [30], the radiated power P_t is calculated as 4.72 dBm. θ_1 ($= 72°$) and ϕ_1 ($= 85°$) are the 3-dB beamwidths in the E- and H-planes, respectively. This corresponds to a dc-to-RF conversion efficiency of 10.5%.

Figure 5.43 shows the radiation patterns measured in the E- and H-planes. The co-polarized fields are at least 20 dB larger than their cross-polarized counterparts in both planes, indicating good linear polarization. Low cross-polarization can be caused by the small circuit size.

5.6   CONCLUSIONS

Recent developments of the reflection-amplifier and coupled-load antenna oscillators have been covered in this chapter. The design methodologies and measurement procedures of the two types of active antennas are discussed, along with several design examples. Despite its various advantages, the performance of the antenna oscillators is limited by the antenna Q factor, which is usually low. Techniques on improving the Q factor have been elaborated. Several successful practical designs of antenna oscillators have also been demonstrated. Finally, some contemporary design issues have been highlighted. With the advent of millimeter-wave and teraheltz eras, antenna oscillators may possibly be used for spatial power combining and other applications.

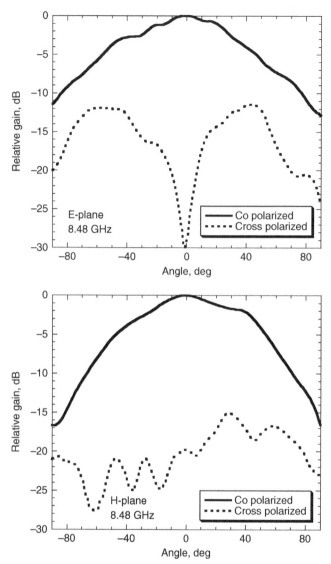

FIGURE 5.43 Normalized radiation measured for the capacitor-load feedback patch antenna in Fig. 5.38. (Courtesy of F. A. Miranda, NASA-Glenn Research Center, Cleveland, Ohio. From [26], copyright © 2008 IEEE, with permission.)

REFERENCES

[1] J. A. Navarro and K. Chang, *Integrated Active Antennas and Spatial Power Combining*. New York: Wiley, 1996.

[2] K. Chang, I. Bahl, and V. Nair, *RF and Microwave Circuit and Component Design for Wireless Systems*. Hoboken, NJ: Wiley, 2002.

[3] A. Mortazawi, T. Itoh, and J. Harvey, *Active Antennas and Quasi-optical Arrays*. Piscataway, NJ: IEEE Press, 1998.

[4] G. D. Vendelin, A. M. Pavio, and U. L. Rohde, *Microwave Circuit Design: Using Linear and Nonlinear Techniques*. Hoboken, NJ: Wiley, 2005.

[5] S. A. Maas, *Nonlinear Microwave and RF Circuits*, 2nd ed. Boston: Artech House, 2003.

[6] G. Gonzalez, *Microwave Transistor Amplifier: Analysis and Design*, 2nd ed. Englewood Cliffs, NJ: Prentice Hall, 1984.

[7] K. Kurokawa, "Some basic characteristics of broadband negative resistance oscillator circuits," *Bell Sys. Tech. J.*, vol. 48, pp. 1937–1955, 1969.

[8] E. N. Ivanov and M. E. Tobar, "Low phase-noise microwave oscillators with interferometric signal processing," *IEEE Trans. Microwave Theory Tech.*, vol. 54, pp. 3284–3294, Aug. 2006.

[9] L. Zhou, Z. Wu, M. Sallin, and J. Everald, "Broad tuning ultra low phase noise dielectric resonator oscillators using SiGe amplifier and ceramic-based resonators," *IET Microwave Antennas Propag.*, vol. 1, no. 5, pp. 1064–1070, Oct. 2007.

[10] Q. Xue, K. M. Shum, and C. H. Chan, "Novel oscillator incorporating a compact microstrip resonant cell," *IEEE Microwave Wireless Compon. Lett.*, vol. 11, pp. 202–204, May 2001.

[11] K. Kawasaki, T. Tanaka, and M. Aikawa, "An octa-push oscillator at V-band," *IEEE Trans. Microwave Theory Tech.*, vol. 58, pp. 1696–1702, July 2010.

[12] S. L. Jang, C. C. Liu, C. Y. Wu, and M. H. Juang, "A 5.6GHz low power balanced VCO in 0.18μm CMOS," *IEEE Microwave Wireless Compon. Lett.*, vol. 19, pp. 233–235, Apr. 2009.

[13] P. Ruippo, T. A. Lehtonen, and N. T. Juang, "An UMTS and GSM low phase noise inductively tuned LC VCO," *IEEE Microwave Wireless Compon. Lett.*, vol. 20, pp. 163–165, Mar. 2010.

[14] F. Giuppi, A. Georgiadis, A. Collado, M. Bozzi, and L. Perregrini, "Tunable SIW cavity backed active antenna oscillator," *Electron. Lett.*, vol. 46, no. 15, pp. 1053–1055, July 2010.

[15] C. H. Mueller, R. Q. Lee, R. R. Romansofsky, C. L. Kory, K. M. Lambert, F. W. V. Keuls, and F. A. Miranda, "Small-size X-band active integrated antenna with feedback loop," *IEEE Trans. Antennas Propag.*, vol. 56, pp. 1236–1241, May 2008.

[16] O. Y. Buslov, A. A. Golovkov, V. N. Keis, A. B. Kozyrev, S. V. Krasilnikov, T. B. Samoilova, A. Y. Shimko, D. S. Ginley, and T. Kaydanova, "Active integrated antenna based on planar dielectric resonator with tuning ferroelectric varactor," *IEEE Trans. Microwave Theory Tech.*, vol. 55, pp. 2951–2956, Dec. 2007.

[17] E. H. Lim and K. W. Leung, "Novel utilization of the dielectric resonator antenna as an oscillator load," *IEEE Trans. Antennas Propag.*, vol. 55, pp. 2686–2691, Oct. 2007.

[18] M. Zheng, Q. Chen, P. S. Hall, and V. F. Fusco, "Oscillator noise reduction in cavity-backed active microstrip patch antenna," *Electron. Lett.*, vol. 33, no. 15, pp. 1276–1277, July 1997.

[19] M. Zheng, P. Gardener, P. S. Hall, Y. Hao, Q. Chen, and V. F. Fusco, "Cavity control of active integrated antenna oscillators," *Microwave Antennas Propag.*, vol. 148, no. 1, pp. 15–20, Feb. 2001.

[20] Y. Chen and Z. Chen, "A dual-gate FET subharmonic injection-locked self-oscillating active integrated antenna for RF transmission," *IEEE Microwave Wireless Compon. Lett.*, vol. 13, pp. 199–201, June 2003.

[21] R. K. Mongia and A. Ittipiboon, "Theoretical and experimental investigations on rectangular dielectric resonator antennas," *IEEE Trans. Antennas Propag.*, vol. 45, pp. 1348–1356, Sept. 1997.

[22] HP85301CK02 Antenna Measurement System: Operating and Service Manual Supplement.

[23] J. Shi, J. X. Chen, and Q. Xue, "A differential voltage-controlled integrated antenna oscillator based on double-sided parallel-strip line," *IEEE Trans. Antennas Propag.*, vol. 56, pp. 2207–2212, Oct. 2008.

[24] K. Chang, K. A. Hummer, and G. K. Gopalakrishnan, "Active radiating element using FET source integrated with microstrip patch antenna," *Electron. Lett.*, vol. 24, no. 21, pp. 1347–1348, Oct. 1988.

[25] I. Waldron, A. Ahmed, and S. Makarov, "Amplifier-Based Active Antenna Oscillator Design at 0.9–1.8 GHz," *IEEE/ACES International Conference on Wireless Communications and Applied Computational Electromagnetics*, Honolulu, HI, Apr. 3–7. 2005, pp. 775–778.

[26] C. H. Mueller, R. Q. Lee, R. R. Romanofsky, C. L. Kory, K. M. Lambert, F. W. Van Keuls, and F. A. Miranda, "Small-size X-band active integrated antenna with feedback loop," *IEEE Trans. Antennas Propag.*, vol. 56, pp. 2686–2691, May 2008.

[27] http://www.cel.com/pdf/datasheets/NE3210S01.pdf.

[28] Hewlett Packard Product Note 11729C-3, A User's Guide for Automatic Phase Noise Measurements, 1986.

[29] C. A. Balanis, *Antenna Theory Analysis and Design*, 2nd ed. New York: Wiley, 1997.

[30] W. L. Stutzman and G. A. Thiele, *Antenna Theory and Design*, 2nd ed. New York: Wiley, 1998.

CHAPTER SIX

Solar-Cell-Integrated Antennas

6.1 INTEGRATION OF ANTENNAS WITH SOLAR CELLS

A solar cell is also known as a photovoltaic or photoelectric cell. It is a device that converts sunlight into electricity. The photovoltaic effect was found by a French physicist, A. E. Becquerel, in 1839, and the phenomenon was later explained by A. Einstein, in 1905. Materials used for modern photovoltaic cells include monocrystalline and polycrystalline silicon, amorphous silicon, cadmium telluride, copper indium selenide/sulfide, and many others. In the past two decades, some work has been reported on the integration of different antennas with solar cell panels [1,10]. Most solar-cell-integrated antennas were used for space-related applications. Figure 6.1 is a photograph showing a satellite bearing antennas and solar cells on different parts of its body. As can be seen from the figure, the two components usually constitute a large surface area. This is very undesirable, as size and weight are among the scarcest resources on a satellite! In 1995, Tanaka et al. [1] covered the resonator and ground surfaces of a microstrip patch antenna with photovoltaic panels. The solar-cell-loaded microstrip antennas were then used in designing a microsatellite. Later, Vaccaro et al. [2,3] proposed etching slot antennas on a solar cell panel, at the expense of reducing the effective illumination area. The configuration of Vaccaro's solar-cell-loaded antenna, which is called Solant, is shown in Fig. 6.2. By cascading six antennas into a 1×6 array, it was demonstrated that solar cells were able to supply a power of 0.821 W with 122 mA at 7 V. An antenna gain of 10 to 11 dBi was also obtained. Shown in Fig 6.3 is the proposed Solant, which was tested for an in-flight space mission [4].

Compact Multifunctional Antennas for Wireless Systems, First Edition. Eng Hock Lim, Kwok Wa Leung.
© 2012 John Wiley & Sons, Inc. Published 2012 by John Wiley & Sons, Inc.

FIGURE 6.1 Antennas and solar cell panels of a satellite.

FIGURE 6.2 Configuration of the Solant. (Courtesy of S. Vaccaro, JAST Antenna Systems, Switzerland. From [2], copyright © 2003 IEEE, with permission.)

As solar-cell-integrated antennas are generally used for outer-space missions, they are required to satisfy the following stringent requirements:

1. *High solar power.* A large photovoltaic area is always required to provide sufficient electricity to power up all the electrical systems of a satellite or spacecraft.
2. *High antenna gain.* High gain is needed to enable electromagnetic signal transmission to points farther in outer space. For example, the orbit of a high-earth-orbit satellite can be as far as 35,786 km from ground [5].

FIGURE 6.3   Solants placed on a satellite. (Courtesy of S. Vaccaro, JAST Antenna Systems, Switzerland. From [4], copyright © 2009 IEEE, with permission.)

3. *Small size and light weight*. Small component size and light weight are requirements to allow rockets, spacecraft, and satellites to escape Earth's gravity.

There are two ways to design solar-cell-integrated antennas. The first is to make use of the solar cell itself as an antenna. As reported by Vaccaro et al. [6], with the simultaneous use of a GaAs solar cell panel itself as an electromagnetic radiator, the antenna gain of the solar-cell-integrated antenna was found to be about 10 dB lower than for the microstrip patch antenna made of copper. Ons et al. [14] suspended a semiconductor-made dipole, itself also a solar cell, at the focus of a solar concentrator designed for collecting solar energy. For the second approach, the antenna and solar cell are made separately but closely integrated into a single module [7]. An effort is required to minimize the interference between the two components. For both methods, however, the antenna gains of the solar-cell-integrated antennas are usually lower than those for their isolated counterparts, probably due to the semiconductor and dielectric losses of the solar cell materials. Some other design issues also need to be looked into. Hwu et al. [8] found that solar cell panels could scatter electromagnetic fields and cause the antenna gain to

TABLE 6.1 Performance Reported for Solar-Cell-Integrated Antennas

	Current (mA)	Voltage (V)	Antenna Gain (dBi)	Antenna Bandwidth (%)
Vaccaro et al. [3]		0.82–0.87	3.4	
Shynu et al. [10]			<2.5	2.42
Lim and Leung [13]	2	2.36	5.3	16.5
Ons et al. [14]	33.2	2.22	11.1	21
Turpin and Baktur [15]			8.04 (directivity)	32

degrade. Tomisawa and Toruda [9] showed that an antenna could cause a neighboring solar cell to radiate ambient signals, introducing additional noise to the wireless system.

Many solar-cell-integrated antennas have been reported in the past few decades, with some of the performances highlighted in Table 6.1. In some studies [2–6], part of the solar cell panel was removed to construct a slot antenna. For other design approaches [10–12], a metallic antenna is designed independently and placed directly on top of a solar cell panel, which also functions as the ground plane of the antenna. In this case, however, the antenna substrate partially blocks the solar cell panel from sunlight reception. To solve this problem, a transparent DRA has been stacked directly on top of a solar cell [13]. Here, the DRA functions simultaneously as a radiating element as well as a focusing lens. Turpin and Baktur [15] integrated meshed patch antennas on solar cell panels for the design of a diversity antenna.

It is always desirable to maximize the surface area of the solar cell so that larger electrical output can be obtained. On the other hand, use of a large number of solar cell panels should be avoided, as it can affect antenna performance. Designing a solar-cell-integrated antenna is sometimes a compromise between the two.

6.2 NONPLANAR SOLAR-CELL-INTEGRATED ANTENNAS

Here, the integration of the dielectric resonator antenna (DRA) and solar cell is explored. In the past two decades, the DRA has been studied extensively because of its small size, low loss, low cost, light weight, and ease of excitation [16,17]. With the use of DR, the antenna size can be usually scaled down by a factor of $\sim 1/\sqrt{\varepsilon_r}$, where ε_r is the dielectric constant of the DR. This feature is very useful for reducing the antenna size. In modern wireless gadgets, compactness is always one of the topmost priorities [18]. This necessitates the development of multifunction components to miniaturize the system.

6.2.1 Solar-Cell-Integrated Hemispherical DRA

It is well known that a lens is a very important component for optical systems. Therefore, it has been of great interest to develop a dual-functional antenna that additionally provides the function of a lens. A dual-functional transparent hemispherical DRA that simultaneously functions as an antenna and a focusing lens for a solar cell was investigated for the first time by Lim and Leung [13]. To make the entire system even more compact, the solar cell is placed beneath the DRA to save the footprint. Here, it is worth mentioning that the DRA can also serve as a protective cover for the solar cell. A conformal strip is used to excite the transparent hemispherical DRA in its dominant mode, TE_{111}. It was found that due to its focusing effect, the transparent hemispherical DR can be used to increase the output voltage and current of the solar-cell-integrated DRA.

Configuration   Figure 6.4 shows the configuration of a dual-functional hemispherical DRA. The transparent DRA [19] was made of borosilicate crown glass, which is commonly known as K9 glass or Pyrex. It has a radius of $R = 28$ mm and is placed above the solar cell. A conformal strip with a width of $w_s = 12$ mm and a length of $l_s = 19$ mm was used to feed the DRA, which was excited in its fundamental mode at 1.87 GHz. The design methods and equations are available in the literature [20–23]. By using an Agilent 85070D dielectric probe kit, the dielectric constant of the glass was measured (Fig. 6.5). It was found that the dielectric constant of Pyrex is $\varepsilon_r \sim 7$ in the frequency range 0.5 to 3 GHz. For verification purposes, the dielectric constants of two well-known materials, air and Teflon, were measured using the same calibrated probe. It was found that the probe was well calibrated, as the measured values were very close to those given by specifications (air, $\varepsilon_r \sim 1$, and Teflon, $\varepsilon_r \sim 2.2$). For all subsequent simulations, the dielectric constant is taken to be $\varepsilon_r \sim 7$ around 1.9 GHz. At optical frequencies, the glass has a much lower dielectric constant of $\varepsilon_r \sim 2.17$, which was calculated from its refractive index of $n = 1.474$ [19,24].

A square solar cell was used to construct the solar-cell-integrated DRA in Fig. 6.4. But since the dielectric properties of the solar cell were not available, the values of $\varepsilon_r = 1.5$ and $\tan \delta = 10$ reported for a solar cell [25] were used in our HFSS simulations. With reference to the figure, the solar cell panel has a side length of $W_c = 55$ mm and a thickness of 1.8 mm. The left and right output pins (for output connection) at the back of the solar cell panel have a height of 0.2 mm. Because of the thickness of the solar cell (1.8 mm) and the height of the output pins (0.2 mm), the DRA has a displacement of 2 mm from the substrate. This information was also used in our HFSS simulations. The substrate has a dielectric constant of $\varepsilon_{rs} = 2.33$, a thickness of $d = 1.57$ mm, and a size of 16×16 cm^2. It serves as an additional insulator between the solar cell and the ground plane. In the experiment, the output pins of the solar cell were connected to a voltmeter and an ammeter for measurements of the voltage and current, respectively. To study

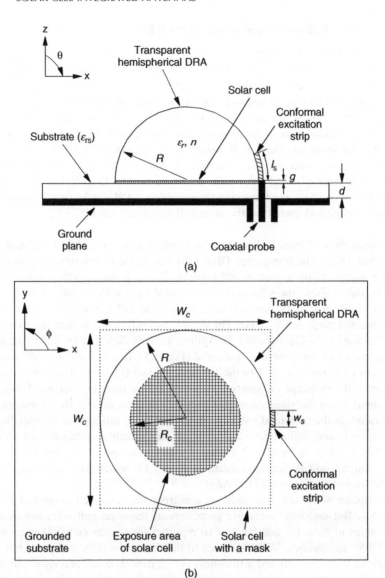

FIGURE 6.4   Dual-functional transparent hemispherical DRA with an underlaid solar cell: (a) front view; (b) top view. (From [13], copyright © 2010 IEEE, with permission.)

the focusing effect of the DRA, the square solar cell is masked with a circular exposure area of radius R_c. A very thin, dark, hard paper was used as the mask and not included in the simulations. Figure 6.6 shows the prototype.

Antenna Performances: Results and Discussion   Ansoft HFSS was used to simulate the antenna part of a solar-cell-integrated DRA, and measurements were carried out using the Agilent 8753 to verify the results. Without considering the

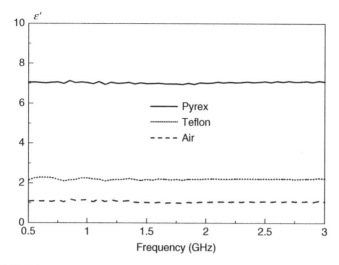

FIGURE 6.5 Dielectric constants measured for Pyrex, air, and Teflon.

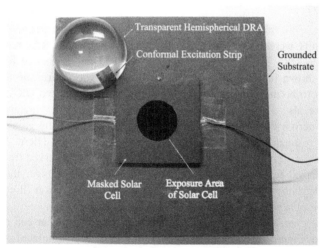

FIGURE 6.6 Transparent hemispherical DRA and solar cell panel. The wires are for the power output of the solar-cell panel. In the actual configuration, the transparent hemispherical DRA is placed on top of the panel. (From [13], copyright © 2010 IEEE, with permission.)

solar cell, the effect of the airgap between the DRA and the substrate is studied first. Figure 6.7 shows the reflection coefficients of the DRA measured and simulated for $g = 0$ and 2 mm, where reasonable agreement between the results is observed for each case. With reference to the figure, the airgap causes the resonance frequency and impedance bandwidth measured ($|S_{11}| \leq -10$ dB) to increase from 1.92 GHz to 2.26 GHz and from 14% to 19%, respectively. The trends are expected because

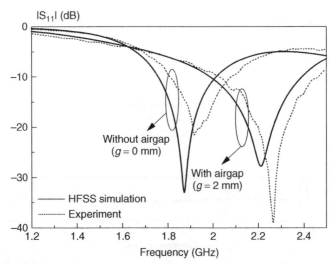

FIGURE 6.7   Reflection coefficients of hemispherical DRAs simulated and measured for $g = 0$ and 2 mm. The results were obtained with no underlaid solar cell. (From [13], copyright © 2010 IEEE, with permission.)

similar results were obtained for the cylindrical and ring DRAs reported in the literature [26–28]. The input impedances simulated and measured for $g = 0$ and 2 mm are shown in Fig. 6.8. As can be seen from the figure, the airgap causes the antenna to be more inductive. The antenna gains of the two DRAs ($g = 0$ and 2 mm) were also measured. As can be seen in Fig. 6.9, they are in the range 4 to 6.8 dBi across their passbands. It is found that the antenna gain is maximum around the resonance for each case, which is typical for the DRA. The radiation patterns for $g = 0$ mm are shown in Fig. 6.10. As can be observed from the figure, the co-polarized fields of both the E- and H-planes are stronger than the cross-polarized fields by more than 20 dB in the bore-sight direction ($\theta = 0°$). Similar results were observed for $g = 2$ mm in Fig. 6.11.

Next, the characteristics of the DRA with the underlaid solar cell (Fig. 6.4) are investigated. Figure 6.12 shows the reflection coefficients simulated and measured for the configuration. As can be observed from the figure, the resonance frequencies of the DRA measured and simulated are 1.94 and 1.89 GHz, respectively, with an error of 2.65%. The impedance bandwidths measured and simulated are given by 16.5% and 22.8%, respectively. Although the DRA in this case also has a displacement of 2 mm from the substrate as for the previous one with the airgap, its measured resonance frequency (1.94 GHz) is lower than that of the airgap case (2.26 GHz). This is because the solar cell increases the effective dielectric constant of the system. It is interesting to note that the resonance frequency measured (1.94 GHz) is quite close to that of the previous DRA (1.92 GHz) resting on the ground plane ($g = 0$ mm). With reference to the figure, a small resonance mode was measured at 2.25 GHz. This mode is caused by the solar cell, which can be

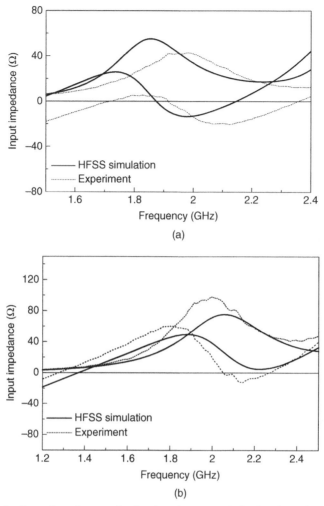

FIGURE 6.8 Input impedances simulated and measured for hemispherical DRAs for (a) $g = 0$ and (b) 2 mm. The results were obtained with no underlaid solar cell.

verified by the fact that it was still observed when a rectangular DRA was used (this will be shown later). The result simulated does not predict this resonance mode, which is not surprising because the exact dielectric parameters of the solar cell are not known.

Figure 6.13 shows the antenna gains of the DRA measured with and without the underlaid solar cell. With reference to the figure, the two antenna gains are very close to each other around the resonance of the DRA. This is a very positive result, as it implies that the loss due to the solar cell is negligibly small. It is observed from the figure that the gain is about 5.3 dBi around the resonance. Figure 6.14 shows the E- and H-plane radiation patterns measured and simulated. As can be

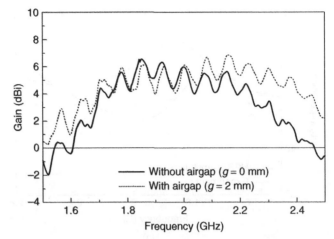

FIGURE 6.9   Antennas gains of hemispherical DRAs measured for $g = 0$ and 2 mm. The results were obtained with no underlaid solar cell. (From [13], copyright © 2010 IEEE, with permission.)

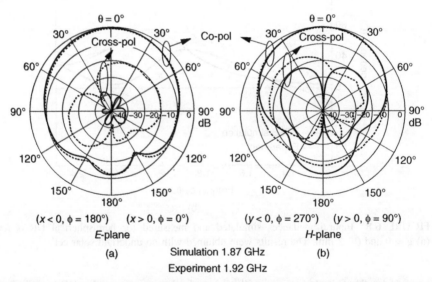

FIGURE 6.10   Normalized radiation patterns of a hemispherical DRA simulated and measured for $g = 0$ mm. There is no underlaid solar cell. (a) E-plane; (b) H-plane. (From [13], copyright © 2010 IEEE, with permission.)

seen from the figure, the co-polarized fields are stronger than their cross-polarized counterparts by at least 22 dB in the bore-sight direction.

Optical Performances: Results and Discussion   As can be seen in Fig. 6.15, a hemispherical lens itself is a good focusing lens for light. The length of the focus can be obtained by $f_i = Rn_2/(n_2 - n_1)$, where R is the center of the sphere, and

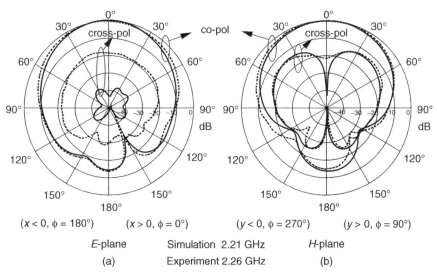

FIGURE 6.11   Normalized radiation patterns of a hemispherical DRA simulated and measured for $g = 2$ mm. There is no underlaid solar cell. (a) E-plane; (b) H-plane. (From [13], copyright © 2010 IEEE, with permission.)

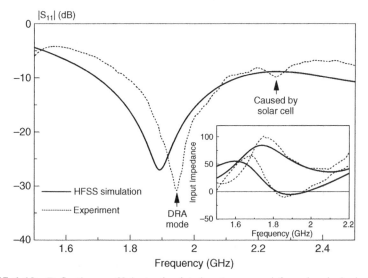

FIGURE 6.12   Reflection coefficients simulated and measured for a hemispherical DRA with an underlaid solar cell (Fig. 6.4). Their corresponding input impedances are shown in the inset. (From [13], copyright © 2010 IEEE, with permission.)

n_1 and n_2 are reflective indexes of the two light-guiding materials [29]. In this case, n_1 is for air and n_2 is for Pyrex. The optical performance of the solar-cell-integrated DRA in Fig. 6.4 is now evaluated. A coherent Sabre Innova argon laser was used in the optical measurements. Parallel blue light beams at a wavelength of 488 nm

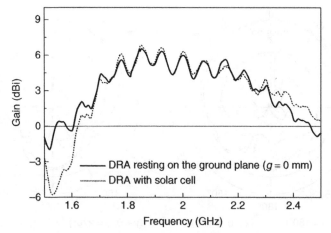

FIGURE 6.13 Antenna gains measured for a hemispherical DRA with and without ($g = 0$ mm) a solar cell. (From [13], copyright © 2010 IEEE, with permission.)

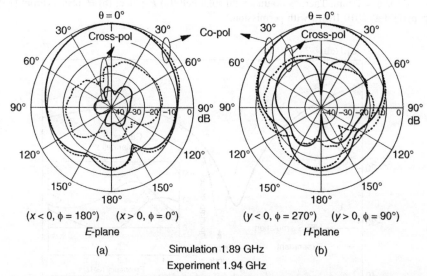

FIGURE 6.14 Normalized radiation patterns simulated and measured for a hemispherical DRA with an underlaid solar cell. (a) E-planel (b) H-plane. (From [13], copyright © 2010 IEEE, with permission.)

were generated. The laser was tuned and provided an even light power of 130 mW to the DRA. To measure the outputs of the solar cell at different illumination angles (θ), the DRA was placed on a rotator, as shown in Fig. 6.16. As can be seen from the figure, a convex lens is used for spreading laser point source into a parallel light beam which is large enough to cover the entire DRA. Figure 6.17 shows the output

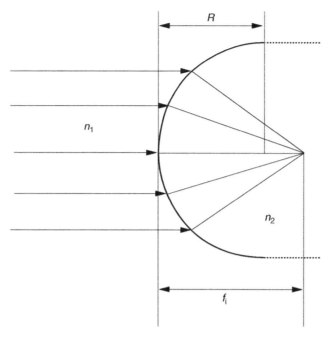

FIGURE 6.15   Focusing length for a hemispherical lens with refractive index n_2.

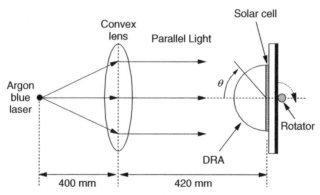

FIGURE 6.16   Top-down view of the coherent Sabre Innova argon laser system for generating a parallel blue light beam. (From [13], copyright © 2010 IEEE, with permission.)

voltage and current of the solar cell measured as a function of the illumination angle for $R_c = 15$ mm. Also shown in the figure are the outputs without the DRA. With reference to the figure, larger outputs can be obtained for $\theta \leq 30°$ by using the DRA because of its focusing effect. With the DRA, the output voltage and current are increased by 13.5% and 27.2% at $\theta = 0°$, respectively. A smaller radius of $R_c = 10$ mm was also used for the mask, and the result is shown in Fig. 6.18. Again, larger outputs were obtained for $\theta \leq 30°$ when the DRA is present. In this

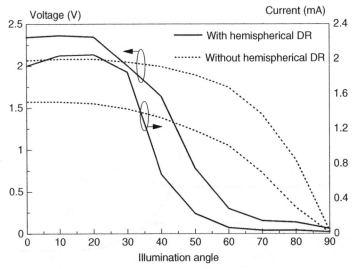

FIGURE 6.17 Output voltages and currents of a solar cell with and without the hemispherical DRA: $R_c = 15$ mm. (From [13], copyright © 2010 IEEE, with permission.)

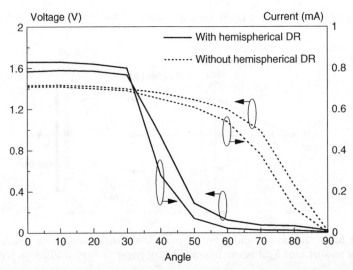

FIGURE 6.18 Output voltages and currents of a solar cell with and without the hemispherical DRA: $R_c = 10$ mm.

case, the voltage and current outputs are increased by 11% and 21.4%, respectively, at $\theta = 0°$ with the use of the DRA. Finally, the voltage and current outputs are measured for a mask with $R_c = 5$ mm and, as can be seen in Fig. 6.19, an increase in the voltage and current outputs of 11.2% and 21.4% is observed for $\theta \leq 25°$. Another observation that can be made from the figures is that a larger illumination

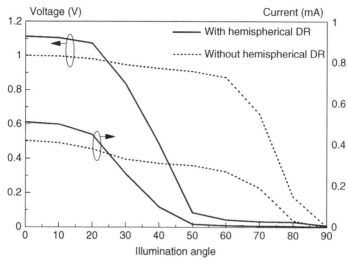

FIGURE 6.19 Output voltages and currents of a solar cell with and without the hemispherical DRA: $R_c = 5$ mm.

area is required to produce higher electricity output. In practical applications, the solar cell panel can be associated with a mechanical rotator so that it can track the light source if needed. In this case, the DRA proposed can be designed into a phased array so that it can scan the electromagnetic beam as the solar cell panel rotates.

Antenna Array Two single-element DRAs (Fig. 6.4) are cascaded to form a 1×2 antenna array, shown in Fig. 6.20. A Krytar MLDD two-way in-phase power divider (Model 6005265 0.5 to 26.5 GHz) is used to feed the two DRAs. The reflection coefficient measured at the input port of the power divider (with the

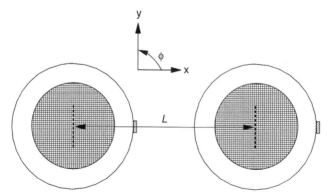

FIGURE 6.20 Top view of a dual-functional transparent hemispherical DRA array with underlaid solar cells.

DRA array connected) is shown in Fig. 6.21. With reference to the figure, the DRA resonance frequency shifts to 1.98 GHz (as compared to 1.94 GHz for the single element). Other resonances at 1.85, 2.08, 2.28, and 2.43 GHz are inherent to the power divider. Next, the antenna gain measured for the DRA array is also given in Fig. 6.22. The maximum antenna gain reads about 8 dBi at 2 GHz. It is about 3 dB higher than the gain for the single-element DRA. Obviously, as can be seen in Fig. 6.23, the array effect is visible in the E-plane. A higher-order array can easily be obtained by cascading more DRA elements.

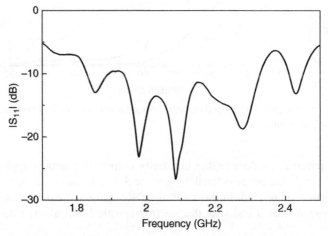

FIGURE 6.21   Reflection coefficient measured for a hemispherical DRA array with underlaid solar cells (Fig. 6.20).

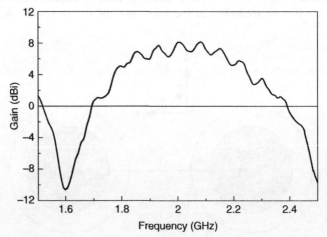

FIGURE 6.22   Antenna gain measured for a hemispherical DRA array with underlaid solar cells.

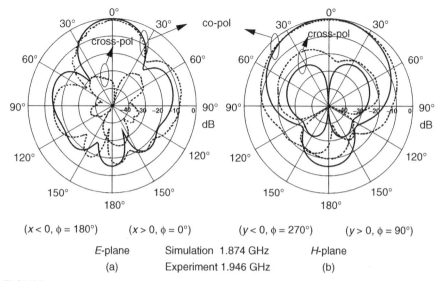

$(x < 0, \phi = 180°)$ $(x > 0, \phi = 0°)$ $(y < 0, \phi = 270°)$ $(y > 0, \phi = 90°)$

E-plane   Simulation 1.874 GHz H-plane

(a)   Experiment 1.946 GHz   (b)

FIGURE 6.23 Normalized radiation patterns simulated and measured for a hemispherical DRA array with underlaid solar cells. (a) E-plane; (b) H-plane.

6.2.2 Solar-Cell-Integrated Rectangular DRA

Configuration In this section the transparent rectangular DRA is also investigated for nonfocusing applications. For ease of comparison, the rectangular DRA was designed to resonate like the hemispherical DRA. The DRA was excited in its fundamental broadside TE_{111} mode using a vertical excitation strip. Figure 6.24 shows a configuration with $\varepsilon_r = 7$, $W = 50$ mm, $H = 22$ mm, $g = 2$ mm, $d = 1.57$ mm, $w_s = 12$ mm, and $l_s = 22$ mm. The same masked solar cell was used again in this part.

Antenna Performances: Results and Discussion The antenna performances of the transparent rectangular solar-cell-integrated DRA are now discussed. Figure 6.25 shows the reflection coefficients that were simulated and measured, with their corresponding input impedances shown in the inset. As can be seen in the figure, the resonance frequencies of the DRA measured and simulated are given by 1.91 and 1.86 GHz, respectively, with an error of 2.7%. The impedance bandwidths measured and simulated are 17.6% and 15.8%, respectively. Again, the resonance due to the solar cell is observed in the measured result. Its resonance frequency shifts slightly from 2.25 GHz to 2.23 GHz due to the changes in the dielectric and excitation-strip loadings. The antenna gain of the transparent rectangular DRA was also measured in Fig. 6.26, which was found to be about 4.2 dBi around the resonance. The radiation patterns simulated and measured are shown in Fig. 6.27. As can be seen in the figure, the cross-polarized fields are weaker than the co-polarized fields by more than 25 dB in the bore-sight direction, showing that the rectangular DRA has very good polarization purity.

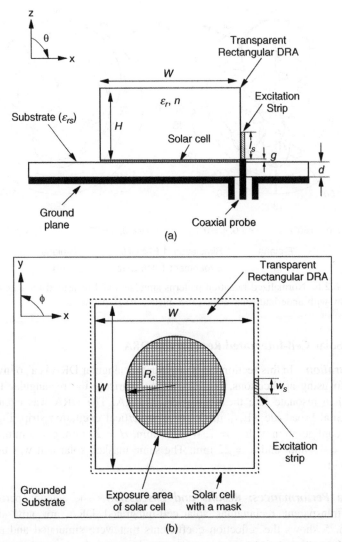

FIGURE 6.24 Transparent rectangular DRA with an underlaid solar cell: (a) front view; (b) top view. (From [13], copyright © 2010 IEEE, with permission.)

Optical Performances: Results and Discussion Figure 6.28 shows output voltages and currents measured using the same masked solar cell with $R_c = 15$ mm. The optical measurement setup shown in Fig. 6.16 is used again for the measurements. With reference to the figure, the rectangular DRA does not increase the outputs of the solar cell, suggesting that the rectangular DRA can be used for applications that do not want the focusing function. From the result, the focusing ability of the hemispherical DRA can be verified. Although the rectangular DRA does not provide a focusing function, its angular range for light reception is much wider

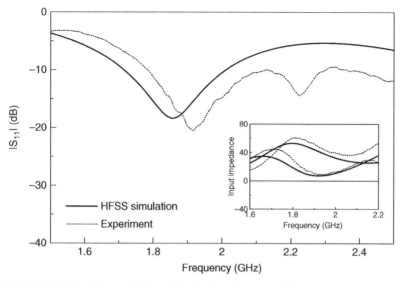

FIGURE 6.25 Reflection coefficients simulated and measured for a transparent rectangular DRA with an underlaid solar cell. Their corresponding input impedances are shown in the inset. (From [13], copyright © 2010 IEEE, with permission.)

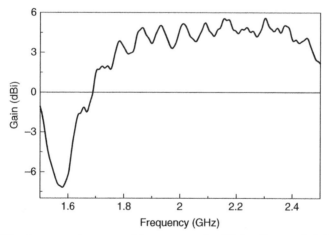

FIGURE 6.26 Antenna gain measured for a rectangular DRA with an underlaid solar cell. (From [13], copyright © 2010 IEEE, with permission.)

than for its hemispherical counterpart. The output voltages and currents measured for $R_c = 10$ mm and 5 mm are also provided in Figs. 6.29 and 6.30, respectively. In general, the outputs do not degrade much for $\theta \le 40°$, implying a larger light reception angle. Again, it is observed that larger illumination area generates higher voltage and current outputs, as expected.

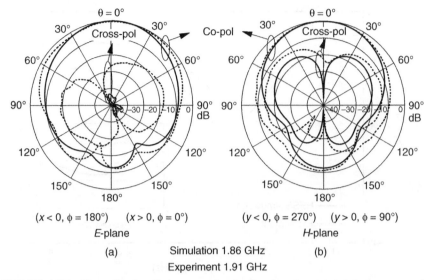

FIGURE 6.27   Normalized radiation patterns simulated and measured for a rectangular DRA with an underlaid solar cell: (a) E-plane; (b) H-plane. (From [13], copyright © 2010 IEEE, with permission.)

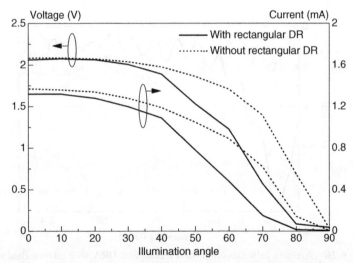

FIGURE 6.28   Output voltages and currents of a solar cell with and without a rectangular DRA: $R_c = 15$ mm. (From [13], copyright © 2010 IEEE, with permission.)

6.3   PLANAR SOLAR-CELL-INTEGRATED ANTENNAS

In solar power plants, solar cells are often combined with light reflectors or concentrators to improve light-power utilization. The light-concentrating solar-cell array, called a *heliostat* [30], is frequently designed to track the movement of the

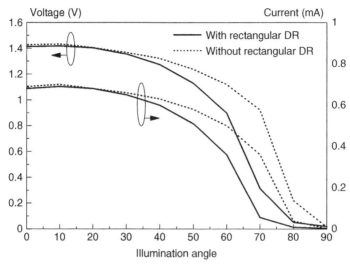

FIGURE 6.29 Output voltages and currents of a solar cell with and without a rectangular DRA: $R_c = 10$ mm.

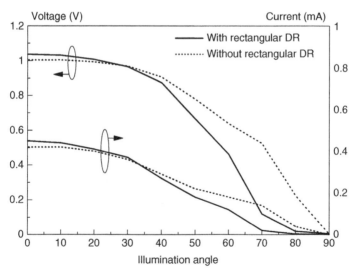

FIGURE 6.30 Output voltages and currents of a solar cell with and without a rectangular DRA: $R_c = 5$ mm.

sun. Traditionally, parabolic disks have been widely used as a solar concentrator to focus sunlight onto solar cells [31]. With the use of solar concentrators, the light intensity at solar cells can be increased and thus a smaller number of photovoltaic cells can be used. In most cases, solar concentrators are made of silicon mirrors. Figure 6.31 shows parabolic and parallel-trench solar concentrators that are used

(a)

(b)

FIGURE 6.31 Two different solar concentrators: (a) parabolic; (b) parallel trench. (Courtesy of S. L. Lau and K. K. Chong, Universiti Tunku Abdul Rahman, Malaysia.)

for heating water. Of course, they can easily be modified into solar power plants by replacing the water containers with solar cells. As the making of a large-aperture parabolic structure can be mechanically challenging, smaller disks are usually cascaded instead to increase the solar power produced [32,33]. Nowadays, various sun-tracking mechanisms [32–34] are deployed to further increase the solar power efficiency of concentrator-type solar plants. Figure 6.32 shows an open-loop

FIGURE 6.32 Open-loop sun-tracking heliostat designed by cascading many parabolic disks, giving a total reflective area of 25 m^2. (Courtesy of K. K. Chong, Universiti Tunku Abdul Rahman, Malaysia.)

sun-tracking system [32] which can provide an output power of 35.7 kW/h per day. It has a total reflective area of 25 m^2. In recent years, to simplify the mechanical design, planar mirrors have been explored as solar concentrators [35]. A solar plant designed by combining 360 pieces of nonimaging planar mirrors is shown in Fig. 6.33. It has a total area of 5760 cm^2 and is able to generate a temperature of 1084°C at the focal point of the parabolic disk. The National Renewable Energy Laboratory in the United States has estimated that a reflector-type power plant would be able to produce electricity at the very low cost of 5.49 cents/kWh by 2020, making solar power one of the cheapest renewable energies in the future [36].

In this section, dual-functional antenna arrays that also serve as light concentrators or reflectors for solar cells are proposed. Since the idea is associated with renewable energy [37], a solar-cell-integrated antenna is also termed a *green antenna*. It will be shown that nonplanar ground planes, which are also light concentrators, are used for green antennas to provide the focusing effect. Instead of using silicon mirrors, metallic plates have been used, as they can easily be designed into electromagnetic radiators.

FIGURE 6.33 Solar plant with nonimaging planar mirrors. It is designed by cascading 360 pieces of planar mirrors, providing a total reflective area of 5760 cm^2. (Courtesy of K. K. Chong, Universiti Tunku Abdul Rahman, Malaysia.)

6.3.1 Solar-Cell-Integrated U-Shaped SPA

Because of their simple structure, wire antennas have been used extensively for many years. They were first utilized by Guglielmo Marconi in 1900 as signal-transmitting radiators. Over the past century, various wire antennas have been explored, such as the monopole, dipole, loop, spiral, and helix antennas. A simple wire antenna has an azimuthally omnidirectional radiation pattern. In 1995, Nakano et al. [38] showed that an L-shaped wire could be used as an efficient excitation probe for C-figured loop antennas. Later, Luk et al. [39] proposed an L-probe-fed suspended plate antenna (SPA). The SPA can provide a very wide impedance bandwidth of more than 35%, with stable radiation patterns across the entire passband. It also provides other advantages, such as low profile, light weight, and ease of excitation and tuning. SPA arrays were designed by Wong et al. [40] to increase the antenna gain. In this section, a U-shaped green antenna is demonstrated. The optical performances were evaluated by Lim et al. [41]. It was found that using the U-shaped ground plane could increase the output voltage of solar cells significantly. Despite the introduction of solar cell panels, the antenna gain of the dual-functional SPA was not much affected.

Configuration Figure 6.34 shows a single-element SPA fed by an L-probe. It will be used to design a solar-cell-integrated green antenna. The SPA was optimized at 2 GHz, with dimensions of $L = 100$ mm, $W = 54$ mm, $H = 18$ mm, $t = 1.5$ mm, $L_h = 21.5$ mm, and $L_v = 11.9$ mm. Next, the configuration of a green antenna is shown in Fig. 6.35. It is a 3×3 array consisting of nine SPA elements placed on a U-shaped ground plane, with parameters of $d = 70$ mm, $W_g = 180$ mm, $L_g = 458$ mm, and $\theta_g = 10°$. The prototype, which is made of aluminum, is shown in Fig. 6.36. As the metallic surface can be scratched easily, such an antenna has to be handled carefully. Two one-sided amorphous solar-cell panels of $W_s = 60$ mm and $L_s = 150$ mm were used in our experiment. The solar cells are cascaded and connected electrically in parallel, as shown in Fig. 6.37. The solar cell panels were fixed using a polyvinyl chloride (PVC) holder at a height of $H = 47$ cm above the ground plane. Both the SPA elements and the ground plane can reflect sunlight to the solar-cell panels and thereby increase the output voltage of the solar system.

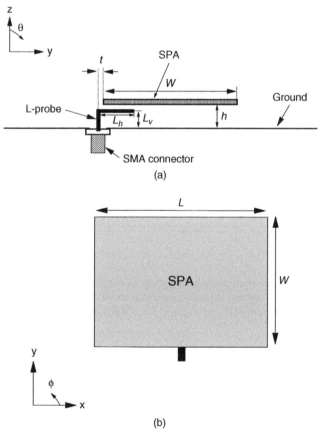

FIGURE 6.34 Single-element L-probe-fed SPA: (a) side view; (b) top view. (From [41], copyright © 2010 IEEE, with permission.)

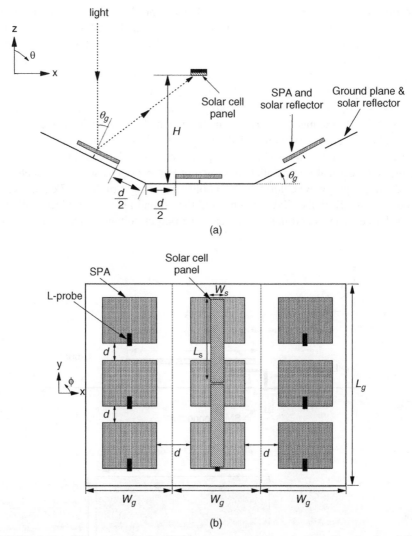

FIGURE 6.35   Solar concentrating green antenna consisting of a light-focusing 3×3 SPA array elements and two solar-cell panels suspended above a U-shaped ground plate: (a) front view; (b) top view. (From [41], copyright © 2010 IEEE, with permission.)

For ease of analyzing the light reflectivity of the ground plane, the suspended plates and solar cell panels are removed. As can be seen from Fig. 6.38, for normal light illumination the solar cells can be suspended at any point in the shaded region, which has a higher light intensity. Multiple solar-cell panels can be used for a larger electricity output. With reference to the figure, at normal light incident, it is observed that the central ground plate does not contribute much to the light beams directed to the solar cells.

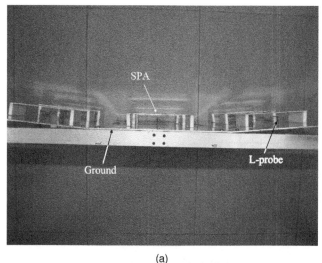

(a)

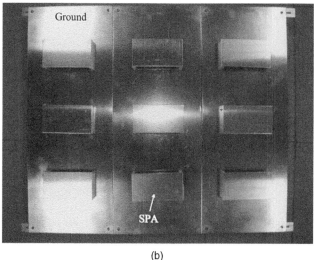

(b)

FIGURE 6.36 Green antenna: (a) front view; (b) top view.

Antenna Performances: Results and Discussion The antenna performance of
a green antenna is discussed next. To begin, a single-element SPA (Fig. 6.34)
resting on a 20×20 cm flat ground plane was optimized using Ansoft HFSS.
Measurements were carried out using an Agilent 8753 network analyzer to verify
the simulations. Figure 6.39 compares simulated and measured input impedances,
where good agreement between the results is observed. The corresponding reflection
coefficients are also shown in the inset. With reference to the inset, the antenna
bandwidths measured and simulated ($|S_{11}| \leq -10$ dB) are about 38% and 42%,

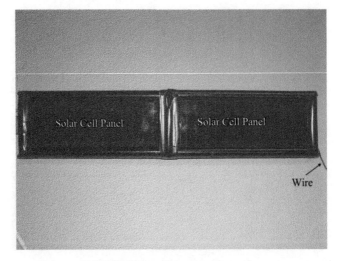

FIGURE 6.37 Cascaded solar cells.

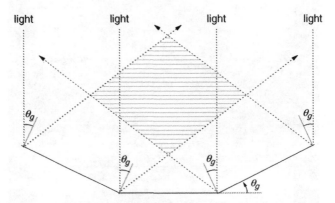

FIGURE 6.38 Analysis of light beams above a U-shaped ground plate.

respectively. Two resonances are observed from the inset. The first resonance is caused by the TM_{01} mode of the suspended plate, which can be verified easily by examining its electric field distribution. The resonance frequencies measured and simulated are 2.12 and 2.09 GHz (1.43% error), respectively. The second resonance is found at around 2.5 GHz. This mode should be caused by the L-probe because its frequency agrees reasonably well with 2.25 GHz, estimated using a primitive formula of $f = c/4(L_v + L_h)$, where c is the speed of light in vacuum. The discrepancy between the two frequencies can be explained by the fact that the primitive formula does not take into account the effect of the suspended plate. In this section our attention is focused on the first (SPA) resonance mode only. Figure 6.40 shows the measured antenna gain of the SPA. With reference to the figure, the gain is about 7 dBi around the SPA mode. This value is typical for

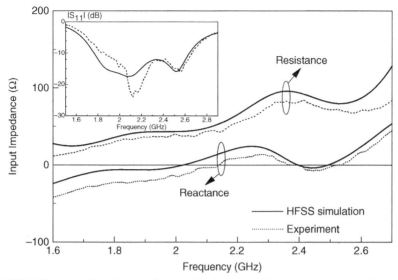

FIGURE 6.39 Input impedances simulated and measured for a single-element SPA. The corresponding reflection coefficients are shown in the inset. (From [41], copyright © 2010 IEEE, with permission.)

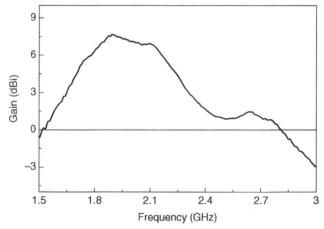

FIGURE 6.40 Antenna gain measured for a single-element SPA. (From [41], copyright © 2010 IEEE, with permission.)

the dominant TM mode of the SPA [39]. Figure 6.41 shows the radiation patterns of the SPA mode. Broadside patterns are found from the figure for both the E- and H-planes, as expected. With reference to the figure, the co-polarized fields are stronger than their cross-polarized counterparts by at least 20 dB in the boresight direction ($\theta = 0°$). In the H-plane, the cross-polarization level increases with θ

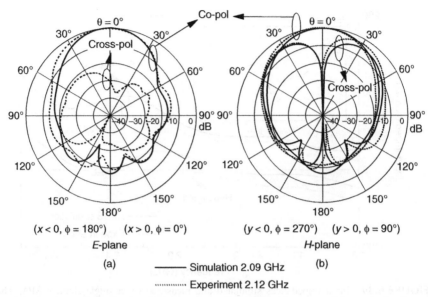

FIGURE 6.41   Normalized radiation patterns simulated and measured for a single-element SPA: (a) E-plane; (b) H-plane. (From [41], copyright © 2010 IEEE, with permission.)

because of the current flowing on the horizontal arm of the L-probe [39]. The simulated E-plane cross-polarized field is too minute to see in the figure.

To feed the SPA array elements of a green antenna, four one-to-three-way Wilkinson power dividers were fabricated and cascaded. Figure 6.42 shows photographs of the Wilkinson power dividers. Four power dividers are cascaded [Fig. 6.42(b)] to feed the nine ports of the SPA array. Its design methodology can be found in [42]. The magnitude passband measured for the power divider is from 1.45 to 2.23 GHz ($|S_{i1}| \leq 4.77 \pm 0.5$ dB, $i = 2, 3$, or 4), giving a bandwidth of 0.78 GHz. The phase bandwidth measured ($|\angle S_{i1} - \angle S_{j1}| < 5°$, $i \neq j$ and i, $j > 1$) is given by 2 GHz (1 to 3 GHz), which is much wider than the magnitude bandwidth. Obviously, the overall bandwidth of the power divider is limited by its magnitude response.

A green antenna is fabricated and tested after the SPA element and Wilkison power divider were designed. It is placed inside a far-field chamber (shown in Fig. 6.43) to measure the antenna gain and patterns. The solar cells are included so that their effects can be analyzed. The antenna gain measured is shown in Fig. 6.44. For ease of comparison, the antenna gain without solar-cell panels is also shown in the figure. As can be seen from Fig. 6.44, the antenna gain measured at 2 GHz is 15.2 dBi, which is only slightly lower than that without the solar-cell panels (15.5 dBi). It shows that the loss introduced by the panels is not very significant. The radiation patterns simulated and measured are shown in Fig. 6.45. It can be observed from the figure that for both E- and H-planes, the co-polarized fields are at least 15 dB stronger than the cross-polarized fields in the bore-sight direction.

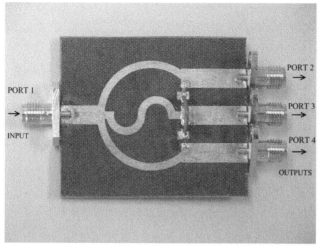

(a)

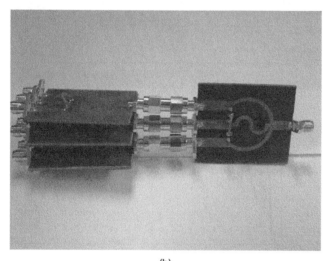

(b)

FIGURE 6.42   One-to-three Wikinson power dividers: (a) single piece; (b) in cascade. (From [41], copyright © 2010 IEEE, with permission.)

The cross-polarized field is stronger in the H-plane. Similar results were obtained for the flat ground plane case [40].

Optical Performances: Results and Discussion   The optical performance of a green antenna is discussed next. It was placed on a rotator to measure the output voltage of the solar-cell panels at different illumination angles (θ). A xenon light source was used in our optical measurements. Figure 6.46 shows the experimental setup. The xenon light source has a wide spectrum and can mimic natural daylight.

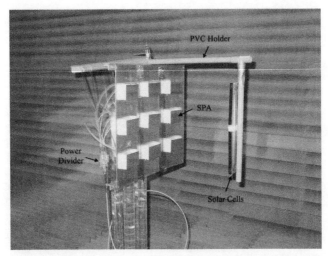

FIGURE 6.43 Green antenna placed in a far-field chamber with the solar cells suspended above it. (Measured at the State Key Laboratory of Millimeter Waves, City University of Hong Kong.)

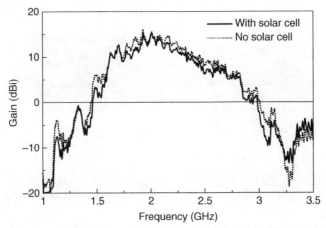

FIGURE 6.44 Antenna gain measured for a green antenna. The antenna gain measured for a green antenna without solar-cell panels is also shown. (From [41], copyright © 2010 IEEE, with permission.)

A parabolic optical dish was used to reflect the light from the source and generate a nearly parallel light beam. Its circular aperture has a diameter of 60 cm. Although an optical lens can produce a much better parallel light beam [13], it is difficult to obtain such a big lens commercially. The height H was found experimentally by

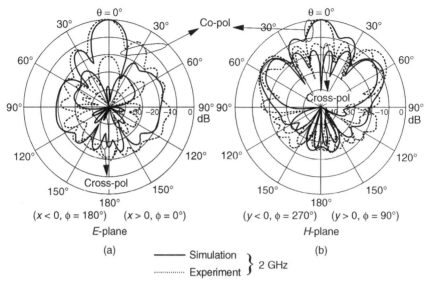

FIGURE 6.45   Normalized radiation patterns simulated and measured for a green antenna: (a) E-planel (b) H-plane. (From [41], copyright © 2010 IEEE, with permission.)

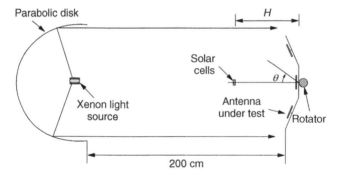

FIGURE 6.46   Top view of an experimental setup for optical measurements. (From [41], copyright © 2010 IEEE, with permission.)

optimizing the voltage output of the solar-cell panels at $\theta = 0°$. Figure 6.47 shows photographs of the optical setup inside an optical dark room and a parabolic optical dish, respectively. Figrue 6.48 shows the output voltages of its solar-cell panels for $\theta_g = 0°$ (flat ground plane) and $10°$ (U-shaped ground plane), with $H = 47$ cm. It can be seen from the figure that when $\theta \leq 22°$, a larger output voltage is obtained using the U-shaped ground plane, due to its focusing effect. When the light is incident normally ($\theta = 0°$), the output voltages are 4.27 and 0.91 V for

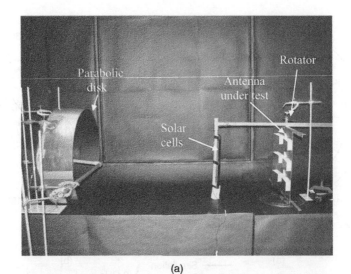

(a)

(b)

FIGURE 6.47 (a) Experimental setup inside an optical dark room. (b) Parabolic optical dish with a xenon light source. (From [41], copyright © 2010 IEEE, with permission.)

the U-shaped and flat ground plane cases, respectively. It was found that an output voltage of 2.38 V is generated when the solar-cell panels are illuminated by the light source directly. In this case, for a fair comparison, the solar-cell panels were placed facing the light beam at a new distance that gives the same total path length (~247 cm). It is interesting to note that the maximum output voltage of a green antenna is found at $\theta = 6°$ instead of $\theta = 0°$. This is caused by the net effect of diffractions, reflections, and light blockage due to the solar-cell panels.

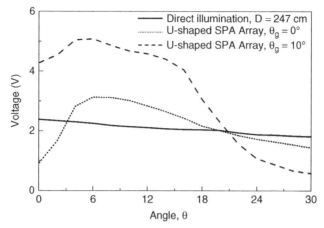

FIGURE 6.48 Output voltages of the solar-cell panels of a Green Antenna versus illumination angles (θ). (From [41], copyright © 2010 IEEE, with permission.)

6.3.2 Solar-Cell-Integrated V-Shaped SPA

In this section, six E-shaped patches are incorporated with a V-shaped ground plane and two solar-cell panels for designing a dual-functional antenna. This type of antenna was first published in a paper by Yang et al. [44]. Like the antenna discussed in the preceding section, the antenna can be used for focusing sunlight onto solar cells.

Configuration Figure 6.49 shows the configuration of a single-element E-shaped patch antenna ($W_1 = 105$ mm, $W_2 = 36$ mm, $W_3 = 21$ mm, $H_1 = 67.5$ mm, $H_2 = 52.5$ mm, $t = 10.5$ mm, and $h = 15$ mm). This antenna is used to design the light-focusing E-shaped patch array shown in Fig. 6.50. As can be seen from the figure, the antenna is a 3×2 array consisting of six E-shaped patches that are evenly placed on a V-shaped ground plane. Other design parameters are $W_g = 162.5$ mm, $L_g = 590$ mm, $W_s = 60$ mm, $L_s = 150$ mm, $H = 55.7$ mm, $d = 94$ mm, and $\theta_g = 5°$. Several Wilkinson power dividers have been designed using the method of Maurin and Wu [42] to feed the antenna array at 1.6 GHz.

Antenna Performances: Results and Discussion Figure 6.51 shows the reflection coefficients simulated and measured for a single-element patch antenna, and their corresponding input impedances are shown in the inset. Two resonant modes were measured at 1.5 GHz (a simulated value of 1.53 GHz) and 1.74 GHz (a simulated value of 1.76 GHz). The impedance bandwidth measured is about 25.3%, which is very close to the simulated value of 25.5%. The antenna gain measured for the light-reflecting E-shaped patch array (Fig. 6.50) is shown in Fig. 6.52. For comparison, the antenna gain for the same array was also measured with the solar cells removed. It can be observed that the solar cells do not introduce much loss

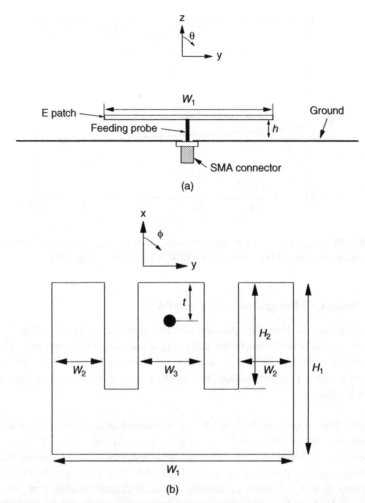

FIGURE 6.49   Single-element E-shaped patch antenna: (a) front view; (b) top view. (From [43], copyright © 2010 IEEE, with permission.)

to the antenna gain at 1.6 GHz. With reference to the figure, the antenna gain for the array with $\theta_g = 0°$ is also given. It is obvious that the solar cells cause the antenna gain of the second mode to degrade. The radiation patterns of the E-shaped patch array ($\theta_g = 5°$ with solar cells) are shown in Fig. 6.53. Broadside patterns are observed, with the co-polarized fields larger than their cross-polarized counterpart by at least 20 dBi in the bore-sight direction.

Optical Performances: Results and Discussion   The same optical measurement setup (shown in Figs. 6.46 and 6.47) is used again for the experiments. It was found that when the light is incident normally ($\theta = 0°$), output voltage with $\theta_g = 5°$ is higher than for $\theta_g = 0°$ by 102% (Fig. 6.54). This shows that the V-shaped patch

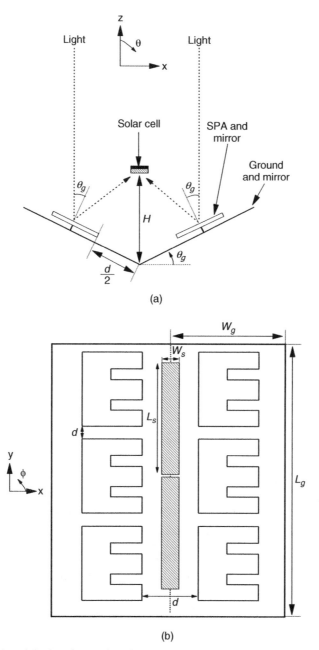

FIGURE 6.50 Light-focusing E-shaped patch array with two solar-cell panels suspended above a V-shaped ground plane: (a) front view; (b) top view. (From [43], copyright © 2010 IEEE, with permission.)

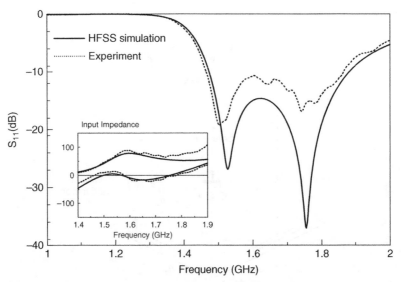

FIGURE 6.51 Reflection coefficients simulated and measured for a single-element E-shaped patch antenna. The corresponding input impedances are shown in the inset. (From [43], copyright © 2010 IEEE, with permission.)

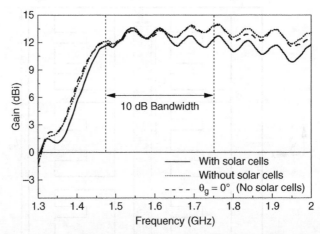

FIGURE 6.52 Antenna gain measured for an E-shaped patch array. The antenna gain measured for the same array without the solar cells is also shown. The results are compared to that for $\theta_g = 0°$. (From [43], copyright © 2010 IEEE, with permission.)

array provides a good focusing effect. Also shown in the figure is the output voltage of the solar cells, being placed at a new distance (255.7 cm) so that it has an optical path identical to that for the inclined array. With reference to the figure, the focusing effect is found for $\theta \leq 13°$ when $\theta_g = 5°$.

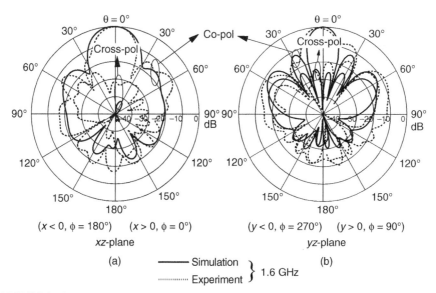

FIGURE 6.53 Normalized radiation patterns simulated and measured for an E-shaped patch array ($\theta_g = 5°$ with solar cells): (a) E-plane; (b) H-plane. (From [43], copyright © 2010 IEEE, with permission.)

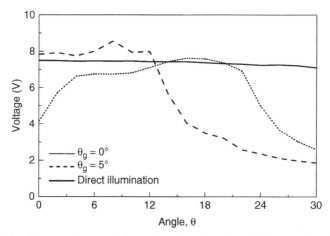

FIGURE 6.54 Output voltages of solar cell panels of a solar-cell-integrated E-shaped patch versus illumination angles (θ). (From [43], copyright © 2010 IEEE, with permission.)

6.4 CONCLUSIONS

Several planar and nonplanar solar-cell-integrated antennas have been discussed in this chapter. Sometimes, designs of such a multifunctional module can be a trade-off between the performances of antenna and solar-cell systems. In the past, solar-cell-integrated antennas were used only in space-related applications. Recently, this technology has quietly come into terrestrial civilian lives [13,45]. By using solar-cell-integrated antennas, wireless systems with self-sustaining power sources can be obtained.

REFERENCES

[1] M. Tanaka, Y. Suzuki, K. Araki, and R. Suzuki, "Microstrip antenna with solar cells for microsatellites," *Electron. Lett.*, vol. 31, no. 1, pp. 5–6, Jan. 1995.

[2] S. Vaccaro, J. R. Mosig, and P. D. Maagt, "Two advanced solar antenna 'Solant' designs for satellite and terrestrial communications," *IEEE Trans. Antennas Propagat.*, vol. 51, pp. 2028–2034, Aug. 2003.

[3] S. Vaccaro, P. Torres, J. R. Mosig, A. Shah, J. F. Zurcher, A. K. Skrivervik, P. D. Maagt, and L. Gerlach, "Stainless steel slot antenna with integrated solar cells," *Electron. Lett.*, vol. 36, no. 25, pp. 2059–2060, Dec. 2000.

[4] S. Vaccaro, C. Pereira, J. R. Mosig, and P. D. Maagt, "In-flight experiment for combined planar antennas and solar cells (Solant)," *IET Microwave Antennas Propag.*, vol. 3, no. 8, pp. 1279–1287, 2009.

[5] http://en.wikipedia.org/wiki/Satellite.

[6] S. Vaccaro, J. R. Mosig, and P. D. Maagt, "Making planar antennas out of solar cells," *Electron. Lett.*, vol. 38, no. 17, pp. 945–947, Aug. 2002.

[7] F. Kustas, "Semiconductor antenna array and solar energy collection assembly for spacecraft," U.S. patent 6,087,991, July 11, 2001.

[8] S. U. Hwu, B. P. Lu, L. A. Johnson, J. S. Fournet, R. J. Panneton, and G. D. Arndt, "Scattering properties of solar panels for antenna pattern analysis," *IEEE International Symposium on Antennas and Propagation Digest*, 1994, vol. 1, pp. 266–269.

[9] M. Tomisawa and M. Tokuda, "Induction characteristics of a solar cell to radiated electromagnetic disturbances," *Asia-Pacific Symposium on Electromagnetic Compatibility and 19th International Zurich Symposium on Electromagnetic Compatibility*, 2008, pp. 538–541.

[10] S. V. Shynu, M. J. R. Ons, P. McEvoy, M. J. Ammann, S. J. McCormack, and B. Norton, "Integration of microstrip patch antenna with polycrystalline silicon solar cell," *IEEE Trans. Antennas Propag.*, vol. 57, pp. 3969–3972, Dec. 2009.

[11] S. V. Shynu, M. J. Ammann, and B. Norton, "Quarter-wave metal plate solar antenna," *Electron. Lett.*, vol. 44, no. 9, pp. 570–571, Apr. 2008

[12] S. V. Shynu, M. J. R. Ons, G. Ruvio, M. J. Ammann, S. McCormack, and B. Norton, "A microstrip printed dipole solar antenna using polycrystalline silicon solar cells," *IEEE International Symposium on Antennas and Propagation and USNC/URSI National Radio Science Meeting*, July 5–11, San Diego, CA, 2008.

[13] E. H. Lim and K. W. Leung, "Transparent dielectric resonator antennas for optical applications," *IEEE Trans. Antennas Propag.*, vol. 58, pp. 1054–1059, Apr. 2010.

[14] M. J. R. Ons, S. V. Shynu, M. J. Ammann, S. J. McCormack, and B. Norton, "Emitter-wrap-through photovoltaic dipole antenna with solar concentrator," *Electron. Lett.*, vol. 45, no. 5, pp. 241–242, Feb. 2009.

[15] T. W. Turpin and R. Baktur, "Meshed patch antennas integrated on solar cells," *IEEE Antennas Wireless Propag. Lett.*, vol. 8, pp. 693–696, June 2009.

[16] K. M. Luk and K. W. Leung, Eds., *Dielectric Resonator Antennas*. London: Research Studies Press, 2003.

[17] A. Petosa, *Dielectric Resonator Antenna Handbook*. Norwood, MA: Artech House, 2007.

[18] A. K. Skrivervik, J. F. Zurcher, O. Staub, and J. R. Mosig, "PCS antenna design: the challenge of miniaturization," *IEEE Antennas Propag. Mag.*, vol. 43, no. 4, pp. 13–27, Aug. 2001.

[19] http://www.camglassblowing.co.uk/gproperties.

[20] K. W. Leung, K. M. Luk, K. Y. A. Lai, and D. Lin, "Theory and experiment of a coaxial probe fed dielectric resonator antenna," *IEEE Trans. Antennas Propag.*, vol. 41, pp. 1390–1398, Oct. 1993.

[21] K. W. Leung, K. M. Luk, K. Y. A. Lai, and D. Lin, "Theory and experiment of an aperture-coupled hemispherical dielectric resonator antenna," *IEEE Trans. Antennas Propag.*, vol. 43, pp. 1192–1198, Nov. 1995.

[22] A. A. Kishk, G. Zhou, and A. W. Glisson, "Analysis of dielectric resonator antennas with emphasis on hemispherical structures," *IEEE Antennas Propag. Mag.*, vol. 36, no. 2, pp. 20–31, Apr. 1994.

[23] K. W. Leung, "Conformal strip excitation of dielectric resonator antenna," *IEEE Trans. Antennas Propag.*, vol. 48, pp. 961–967, June 2000.

[24] Datasheet: Borosilicate Glass Properties.

[25] J. Dheepa, R. Sathyamoorthy, A. Subbarayan, S. Velumani, P. J. Sebastian, and R. Perez, "Dielectric properties of vacuum deposited Bi2Te3 thin films," *Solar Energy Mater. Solar Cells*, vol. 88, no. 2, pp. 187–198, July 2005.

[26] G. P. Junker, A. A. Kishk, A. W. Glisson, and D. Kajfez, "Effect of an air gap around the coaxial probe exciting a cylindrical dielectric resonator antenna," *Electron. Lett.*, vol. 30, no. 3, pp. 177–178, 1994.

[27] G. P. Junker, A. A. Kishk, A. W. Glisson, and D. Kajfez, "Effect of an air gap on a cylindrical dielectric resonator antennas operating in the TM_{01} mode," *Electron. Lett.*, vol. 30, no. 2, pp. 97–98, 1994.

[28] S. M. Shum and K. M. Luk, "Characteristics of dielectric ring resonator antenna with an air gap," *Electron. Lett.*, vol. 30, pp. 277–278, Feb. 1994.

[29] E. Hecht, *Optics*, 4th ed. Reading, MA: Addison Wesley, 2002.

[30] W. B. Stine and R. W. Halligan, Eds, *Solar Energy Fundamentals and Design with Computer Applications*. New York: Wiley, 1985, pp. 135–262.

[31] R. B. Diver and T. A. Moss, "Practical field alignment of parabolic trough solar concentrators," *J. Solar Energy Eng.*, vol. 129, pp. 153–159, May 2007.

[32] K. K. Chong, C. Wong, F. Siaw, T. Yew, S. Ng, M. Liang, Y. Lim, and S. Lau, "Integration of an on-axis general sun-tracking formula in the algorithm of an open-loop sun-tracking system," *Sensors*, vol. 9, pp. 7849–7865, Sept. 2009.

[33] K. K. Chong, "Optimization of nonimaging focusing heliostat in dynamic correction of astigmatism for a wide range of incidence angles," *Opt. Lett.*, vol. 35, no. 10, pp. 1614–1616, 2010.

[34] K. K. Chong, "Optical analysis for simplified astigmatic correction of non-imaging focusing heliostat," *Solar Energy*, vol. 84, pp. 1356–1365, 2010.

[35] K. K. Chong, F. Siaw, C. Wong, and G. Wong, "Design and construction of non-imaging planar concentrator for concentrator photovoltaic system," *Renewable Energy*, vol. 34, pp. 1364–1370, 2009.

[36] Report: Assessment of Parabolic Trough and Power Tower Solar Technology Cost and Performance Forecasts, NREL/SR-550-34440, Oct. 2003.

[37] http://en.wikipedia.org/wiki/Sustainable_energy#Green_energy.

[38] H. Nakano, H. Yoshisda, and Y. Wu, "C-figured loop antennas," *Electron. Lett.*, vol. 31, no. 9, pp. 693–694, Apr. 1995.

[39] K. M. Luk, C. L. Mak, Y. L. Chow, and K. F. Lee, "Broadband microstrip patch antenna," *Electron. Lett.*, vol. 34, no. 15, pp. 1442–1443, July 1998.

[40] H. Wong, K. L. Lau, and K. W. Luk, "Design of dual-polarized L-probe patch antenna arrays with high isolation," *IEEE Trans. Antennas Propag.*, vol. 52, pp. 45–52, Jan. 2004.

[41] E. H. Lim, K. W. Leung, S. C. Su, and H. Y. Wong, "Green antenna for solar energy collection," *IEEE Antennas Wireless Propag. Lett.*, vol. 9, pp. 689–692, July 2010.

[42] D. Maurin and K. Wu, "A compact 1.7–2.1GHz three-way power combiner using microstrip technology with better than 93.8% combining efficiency," *IEEE Microwave Guided Wave Lett.*, vol. 6, pp. 106–109, Feb. 1996.

[43] E. H. Lim, K. W. Leung, G. H. Khor, and K. K. Chan, "A solar power plant with light-reflecting E-shaped patch antenna," *IEEE International Symposium on Antennas and Propagation and USNC/URSI National Radio Science Meeting 2010*, Toronto, Oatario, Canada, July 11–17, 2010.

[44] F. Yang, X. Zhang, X. Ye, and Y. Rahmat-Samii, "Wide-band E-shaped patch antennas for wireless communications," *IEEE Trans. Antennas Propag.*, vol. 49, pp. 1094–1100, July 2001.

[45] T. Schuetze, H. Hullmann, and C. Bendel, "Multifunctional photovoltaic building design and efficient photovoltaic use of solar energy," *World Climate & Energy Event*, Rio de Janeiro, Brazil, Mar. 17–19, 2009.

Index

3-dB bandwidth, 62

airgap, 191
amorphous silicon, 185
amplifier
  low-noise, 5, 77, 117–18, 123, 131
  power, 4, 117
analog modulation, 3
anechoic chamber, 125, 127
antenna
  bandwidth, 8–9
  dielectric resonator, 32, 188
  directional, 12
  E-shaped patch, 219
  fractal patch, 103
  gain, 12
  leaky-wave, 8, 105
  nonplanar, 7, 16
  omnidirectional, 12
  oscillating, 145
  planar, 7, 14
  quasi-Yagi, 161
  reconfigurable, 85
  slot, 54, 89
  solar-cell-integrated, 185, 188,
    204
  surface-wave, 8
  suspended plate, 208
  wearable, 118
  wire, 208
  Yagi, 91
  Yagi-Uda, 109
antenna array, 199, 210, 221
antenna filter
  configuration, 29
  dielectric resonator, 31
  microstrip-based, 50
  ring-patch, 50, 52
  ring-slot, 54
  ultrawideband, 57

antennafier, 117
antennamitter, 117
antenna-in-package, 78
antenna-on-package, 78
antenna oscillator
  coupled-load, 149, 167
  coupled-load microstrip patch, 167
  dielectric resonator, 151–2, 158
  differential planar, 161
  microstrip patch, 151, 171
  reflection amplifier, 149, 152
  slot, 151
  substrate integrated waveguide, 151
  voltage-controlled integrated, 161
antenna package, 67
argon laser, 195
axial ratio, 13

balun filter
  magnetic-coupled, 64
  multiband, 60, 62
  patch, 65
  ring, 60
  single-band, 60, 62
Barkhausen criterion, 147, 170
baseband modules, 3
biasing scheme
  common-base, 149
  common-emitter, 150
  common-source, 119
bidirectional radiation, 59
bilateral transceiver, 5
bits
  channel, 2
  source, 2
block diagram, 146
borosilicate crown glass, 189

cadmium telluride, 185
calibration, 154

Compact Multifunctional Antennas for Wireless Systems, First Edition. Eng Hock Lim, Kwok Wa Leung.
© 2012 John Wiley & Sons, Inc. Published 2012 by John Wiley & Sons, Inc.

WILEY SERIES IN MICROWAVE AND OPTICAL ENGINEERING

KAI CHANG, Editor
Texas A&M University

Printed and bound by CPI Group (UK) Ltd, Croydon, CR0 4YY

16/04/2025

14658603-0001